The Biology of Marine Plants

M. J. Dring

Department of Botany,
Queen's University of Belfast

CAMBRIDGE
UNIVERSITY PRESS

PUBLISHED BY THE PRESS SYNDICATE OF THE UNIVERSITY OF CAMBRIDGE
The Pitt Building, Trumpington Street, Cambridge CB2 1RP, United Kingdom

CAMBRIDGE UNIVERSITY PRESS
The Edinburgh Building, Cambridge CB2 2RU, UK http://www.cup.cam.ac.uk
40 West 20th Street, New York, NY 10011–4211, USA http://www.cup.org
10 Stamford Road, Oakleigh, Melbourne 3166, Australia

First published by Edward Arnold 1982
First published by Cambridge University Press 1992
Reprinted 1994, 1996, 1998

Printed in the United Kingdom at the University Press, Cambridge

A catalogue record for this book is available from the British Library

ISBN 0 521 42765 7 paperback

Contents

Preface

Marine plants have rarely been discussed as a distinct and self-contained group. They have traditionally been treated either as the poor relations of marine animals in courses and texts on Marine Biology, or as examples of particular groups of algae, where the essential 'marine-ness' of marine plants tends to disappear amongst the taxonomic and morphological parallels with freshwater algae. This book attempts to liberate marine plants from both of these traps by providing an introduction to recent analytical and experimental studies of plant growth in the sea. The physics and chemistry of the marine environment are examined with specific reference to the requirements of marine plants, and most of the book concentrates on those aspects of physiology which are unique to marine plants, or which help us to understand their ecology. This discussion emphasizes the importance of a good background knowledge of the environment for critical measurements of the most important factors, and the necessity for experimental work on marine plants to be well quantified, and to be conducted in ecologically relevant conditions.

Since over 90% of the species of marine plants are algae, most of the book is devoted to the marine representatives of this group, with examples from all oceans and coasts of the world, but there is no detailed morphological or taxonomic treatment. Phytoplankton and seaweeds are discussed together as far as possible, in spite of the obvious morphological and ecological contrasts between them, in order to obtain an integrated picture of the biology of marine plants in general. There is, however, a deliberate bias in certain chapters towards the seaweeds, since the physiology and ecology of these plants has not been so fully covered in books at this level as has the ecology of phytoplankton. Marine angiosperms are also discussed where appropriate alongside the autotrophic algae, and the ecological roles of bacteria and fungi in the sea are covered in a separate, final chapter.

A basic knowledge of plant biology — and especially plant physiology — is assumed, but sufficient background is provided for each topic to remind readers of the ideas and terms referred to in the text. It is hoped that the book will provide a stimulating and challenging text for second and third year undergraduate courses, as well as being of interest and value to postgraduate students and specialists in other areas of marine biology.

Acknowledgements

Most of the line drawings in this book are adaptations of previously published illustrations, and I wish to thank all authors and publishers for their permission to reproduce such material. Original prints of photographs were kindly provided by Robert Buggeln, Klaus Lüning, Betty Moss and Susan Waaland. Professor A.J. Willis has read every page of the manuscript with great care, and has provided valuable guidance and encouragement throughout the preparation of the book. Numerous other colleagues have kindly read and commented on parts of the manuscript, and I am particularly grateful for the detailed and constructive criticisms of Klaus Lüning, Graham Savidge, David Jewson and Philip Grime. Since I have not always accepted their advice, however, they are in no way responsible for the faults and errors that may remain.

1982   M.J.D.

1

Marine Plants: taxonomic, morphological and ecological categories

It is generally accepted that life on earth began in the sea and that, until about 450 million years ago, all plants were marine plants. The next 400 million years witnessed the evolution of the land flora, including bryophytes, pteridophytes, gymnosperms and flowering plants. The story of this invasion of the land is usually told in terms of the developments in morphology and reproduction that made plants less and less dependent on the presence of free water, but one aspect that is often overlooked is that, in this process, plants lost their ability to live in sea water. Throughout the modern bryophytes, pteridophytes and gymnosperms, there is not a single marine species and, in the angiosperms, which have evolved more than 200 000 species adapted to almost every terrestrial and freshwater habitat, there is only one small group of seagrasses (about 50 species world-wide[89]) which can be described as truly marine. Even the fungi are poorly represented in the sea, since marine species account for only 1% of the total fungal flora.[125] Thus the sea remains — as it must have been in pre-Devonian times — the province of the algae, and today about 90% of all the species of marine plants belong to one or other of the groups of algae (Table 1.1).

TAXONOMIC CLASSIFICATION

Many of these algal groups are now themselves represented mainly by freshwater species, and only two groups — the brown (Phaeophyta) and the red algae (Rhodophyta) which predominate among attached marine plants — remain almost exclusively marine. These two groups are joined on the sea floor by a few distinct groups of macroscopic green algae (Chlorophyta: the Ulvales and the siphonaceous orders, see p. 8), which are again almost exclusively marine, in contrast to the microscopic green algae, which are predominantly freshwater or sub-aerial forms. The microscopic marine flora is dominated by two groups of brown-coloured algae, the diatoms (Chrysophyta) and the dinoflagellates (Pyrrhophyta). Of these two groups, the dinoflagellates are again predominantly marine, but the diatoms are more evenly divided, in terms of numbers of species, between marine and freshwater habitats (Table 1.1).

When considering the representation of different plant groups in the marine flora in terms of numbers of species, it is important to remember that species vary enormously in numbers of individuals and in the size of those individuals, and that the number of species described in any one group probably depends more on the amount of study that the group has received than on the actual diversity within the group. The absolute numbers listed in Table 1.1 should, therefore, be treated with caution, and the values for some of the groups could possibly be increased by 50 or even 100%. For example, some estimates place the total number of species of diatoms (which account for over 90% of the Chrysophyta in Table 1.1) at 12 000, of red algae at 3900, and of brown algae at 1500. Few of these recent estimates separate the numbers of marine and freshwater species, but it is probably reasonable to assume that the numbers of marine species will increase roughly in proportion to the increase in total numbers in any one group. There will probably, therefore, be little overall change in the percentage representation of different groups in the marine flora, and this approach to plant classification provides a clear indication of which groups are predominantly marine, and which are mainly or entirely confined to freshwater and terrestrial habitats.

The poor representation of so many major plant groups in the marine flora, and the virtual restriction of other groups to the sea, suggests that growth in the sea creates problems for plants that are quite different from those created by growth in fresh water or on land (where soil water is also 'fresh'). Few groups of plants have evolved that seem able to cope equally well with both environments — the diatoms being perhaps the only example (see Table 1.1). A major objective of the study of marine plants must clearly be to identify the critical differences between the sea and freshwater as an environment for plant growth.

Three of the groups of marine algae that have been mentioned so far — the brown, the red and the green algae — have been described simply on the basis of their colour. The colour of the thallus was introduced as a taxonomic character for the algae as long ago as 1836, and the modern classification of the algae still uses pigment composition as a primary character (Table 1.2). The 'green algae' (including Chlorophyta, Charophyta, Euglenophyta) have a similar pigmentation to the higher plants, and contain chlorophylls a and b and β-carotene. The remaining groups all contain chlorophyll a and β-carotene, but lack chlorophyll b. The brown-coloured algae (Phaeophyta, Chrysophyta, Pyrrhophyta and Cryptophyta) all contain chlorophyll c in addition to chlorophyll a, but derive their characteristic brown colour from a mixture of xanthophyll pigments, of which fucoxanthin is the most important in the Phaeophyta and the Chrysophyta, while peridinin is typical of dinoflagellates. Cryptomonads may be red, blue or green in colour, depending on the culture conditions, because of the

Table 1.1   Approximate number of marine species in each major plant group.

Plant group	Total species	Marine species	Marine spp. as percent of group	Contribution of group to marine flora (%)
Algae:	19300	8462	43.8	93.8
Chlorophyta	6500	900	13.8	10.0
Charophyta	250	0	0.0	—
Euglenophyta	450	15	3.3	0.2
Phaeophyta	1000	997	99.7	11.0
Chrysophyta	6000	3000	50.0	33.2
Pyrrhophyta	1000	900	90.0	10.0
Cryptophyta	100	50	50.0	0.6
Rhodophyta	2500	2450	98.0	27.1
Cyanophyta	1500	150	10.0	1.7
Fungi:	50000	500	1.0	5.5
Myxomycetes	500	0	0.0	—
Phycomycetes	1500	100	6.7	1.1
Ascomycetes	20000	210	1.1	2.3
Basidiomycetes	16000	10	0.06	0.1
Deuteromycetes	12000	180	1.5	2.0
Lichenes	15500	15	0.1	0.2
Bryophyta	22000	0	0.0	—
Pteridophyta	11000	0	0.0	—
Spermatophyta	220700	49	0.02	0.5
Gymnospermae	700	0	0.0	—
Angiospermae	220000	49	0.02	0.5
Monocotyledones	50000	49	0.10	0.5
Dicotyledones	170000	0	0.0	—
TOTAL	338500	9026	2.67	

Main sources: Smith, G.M. (1955) *Cryptogamic Botany*, Vol. 1, 2nd edition. McGraw-Hill, New York; Altman, P.L. and Dittmer, D.S. (1972) *Biology Data Book*, Vol. 1, 2nd edition. Federation of American Societies for Experimental Biology.

presence of phycobilin pigments, and these are also the major accessory pigments in red and blue-green algae. There are two basic types of phycobilin, and most species in these groups contain some of each. The red pigment, phycoerythrin, predominates in most red algae (Rhodophyta), and the blue-green pigment, phycocyanin, predominates in typical blue-green algae (Cyanophyta), but the balance between the pigments varies considerably within each group, so that some 'blue-green' algae (e.g. *Phormidium*) may be bright red in colour, whereas some 'red' algae may appear brown or olive-green (e.g. *Porphyra* spp.). The physiological and

Table 1.2 Primary classification of algae.

Division	Popular name	Major accessory pigments	Storage products	No., arrangement, type of flagella	Major marine representatives
Chlorophyta	Green algae	Chlorophyll *b*	Starch (amylose + amylopectin)	2, 4 or many equal, anterior, smooth	Siphonaceous orders, Ulvales
Charophyta	Charophytes	(as Chlorophyta, but distinguished by vegetative and reproductive morphology)		anterior, smooth	No marine representatives
Euglenophyta	Euglenoids	Chlorophyll *b*	Paramylon	1 anterior, tinsel	Few marine representatives
Phaeophyta	Brown algae	Chlorophyll c_1 + c_2, fucoxanthin	Laminaran	2 unequal, lateral, smooth + tinsel	Whole group marine
Chrysophyta	Yellow-brown or golden-brown algae	Chlorophyll c_1 + c_2, fucoxanthin	Chrysolaminaran	1–3 anterior, various	Diatoms, coccolithophorids, silicoflagellates
Pyrrhophyta	Dinoflagellates	Chlorophyll c_2, peridinin	Starch	2 equal, smooth	Dinoflagellates
Cryptophyta	Cryptomonads	Chlorophyll c_2, phycobilins	Starch	2 equal, lateral, both tinsel	Some cryptomonads
Rhodophyta	Red algae	Phycoerythrin ± phycocyanin	Floridean starch (amylopectin)	None	Nearly all species marine
Cyanophyta	Blue-green algae	Phycocyanin ± phycoerythrin	Myxophycean starch (glycogen-like)	None	Planktonic and benthic filaments

ecological implications of these variations in the pigment composition of marine algae are considered in Chapter 3.

Two other important characters in the primary classification of algae are the chemical nature of the storage products,[46] and the number, type and arrangement of the flagella in motile cells[19, 33] (Table 1.2). The three forms of starch that are found in the algae differ in the degree of branching and in the size of the polysaccharide molecule, but all three consist of glucose units linked in the same way as in higher plant starch (i.e. α-(1,4) linkages). The other storage products produced by algae (laminaran, chrysolaminaran and paramylon) also consist of glucose units, but are polymerized by β-(1,3) linkages. Almost all of these storage polysaccharides are synthesized by at least some marine species, and by some freshwater species. There seems to be no particular relationship, therefore, between the type of storage product and the marine habit.

With the important exception of the red and blue-green algae, all of the algal groups contain species which are flagellated for at least part of their life history. The number of flagella ranges from one (e.g. male gametes of centric diatoms, silicoflagellates) to many (e.g. zoospores of the siphonaceous green alga *Derbesia*), but is most commonly two, and these may be either equal or unequal in length. The flagella may be inserted at the anterior apex of the cell, as in most green algae and chrysophytes, or they may be inserted behind the apex or at the side of the cell (e.g. motile cells of brown algae, dinoflagellates and cryptomonads). The dinoflagellates are characterized by a flagellar arrangement that is peculiar to the group. Two flagella arise from the side of the cell, and one encircles the cell while the other is directed backwards from the point of insertion. Two distinct types of flagellum may occur: a smooth, 'whiplash' flagellum, and a 'flimmer' or 'tinsel' type which is covered in a series of short hairs. Almost all of the different combinations of these characters found among the algae can also be found in at least some marine species, but most also occur in freshwater and sub-aerial algae. The only arrangement that is really restricted to marine plants is that of the motile cells of brown algae, with two unequal flagella, one of each type, inserted at the side of a kidney-shaped cell. It is difficult to believe, however, that this arrangement imposes any particular disadvantage in fresh water, and its restriction to sea water is probably an evolutionary accident. The physiological and ecological characteristics of the vegetative thalli of brown algae seem more likely to be responsible for the restriction of these plants to the sea than does the flagellar arrangement of their reproductive cells.

MORPHOLOGICAL CLASSIFICATION

The primary classification of the algae presented in Table 1.2 aims to express the fundamental evolutionary or phylogenetic relationships between the different species, but it is possible, and often useful, to classify organisms according to different criteria, and for different purposes. Table 1.3 presents a morphological classification of marine plants (excluding fungi,

Table 1.3 Morphological classification of marine plants, with important or familiar examples of each category.

Type of thallus		CHLOROPHYTA	PYRRHOPHYTA	RHODOPHYTA	CHRYSOPHYTA
Unicells:	Flagellate	*Dunaliella, Chlamydomonas*	Dinoflagellates (cryptomonads) 'Marine amoebae'	—	Coccolithophorids, silicoflagellates 'Marine amoebae' Diatoms
	Rhizopodial	—	'Marine amoebae'	(some spores)	
	Protococcoidal	*Chlorella marina*	—	*Porphyridium*	Diatoms
Colonies:	Amorphous	(Mostly freshwater)	*Ceratium*	—	Colonial diatoms, *Phaeocystis*
	Coenobial	(Entirely freshwater)	—	—	—
Filaments:	Unbranched	*Ulothrix, Stichococcus*	CYANOPHYTA *Oscillatoria, Lyngbya Phormidium, Calothrix*		*Diatoms (e.g. Melosira, Chaetoceros)*
	Branched	*Cladophora*	*Brachytrichia*	*Porphyra* (conchocelis)	—
	Heterotrichous Crustose	*Phaeophila Ulvella*	PHAEOPHYTA *Ectocarpus Ralfsia*	*Rhodochorton Lithothamnion, Petrocelis*	—
	Pseudo-parenchymatous	—	*Desmarestia, Leathesia*	*Chondrus, Eucheuma, Gigartina, Palmaria*	—
Coenocytes:	Simple Uniaxial	*Valonia, Acetabularia Derbesia, Caulerpa, Bryopsis*	—	—	*Vaucheria*
	Multiaxial	*Codium, Udotea, Halimeda*	—	—	—
Parenchymatous thalli	2-dimensional	*Ulva, Monostroma, Enteromorpha*	*Scytosiphon, Petalonia*	*Porphyra, Delesseria*	—
	3-dimensional		*Dictyota, Fucus, Laminaria*	—	—
	Differentiated	SPERMATOPHYTA Sea grasses	—	—	—

which are discussed in Chapter 9), and lists the most important and familiar examples of each category. The taxonomic position of each example is also indicated in order to show the relationship between this morphological classification and the primary taxonomy of the algae. It is clear that a substantial amount of parallel evolution must have occurred within the different groups of algae, since most of the morphological types are represented by algae from several different groups. Most of the different types of *flagellate unicells* (see above) are represented in the marine flora, but the dinoflagellates and the coccolithophorids are the most abundant. The latter group is named after the elaborate and varied scales, made of calcium carbonate (the 'coccoliths'), that totally cover the cell surface. The fine detail of the coccoliths is well preserved in marine sediments, and fossil forms dating back to the early Jurassic have great stratigraphic value (see p. 144), as well as testifying to the importance of these plants in the marine ecosystem. The fossil record is also rich in the siliceous remains of silicoflagellates, although these are rare in modern seas. Unicells exhibiting amoeboid movement are described as *rhizopodial* and a few species of 'marine amoebae', with or without chloroplasts, are attributed to the Pyrrhophyta and the Chrysophyta. The most important unicellular plants in the sea, however, are the diatoms, which are classed — together with the familiar and ubiquitous green alga *Chlorella* — as *protococcoidal*. In these plants, the vegetative cells have no specialized organs for movement, such as flagella or pseudopodia, but they are not necessarily immobile. Benthic diatoms are often very active, gliding along a mucilage trail secreted from the underside of the cell.

In many species of diatoms, the daughter cells fail to separate completely after cell division; in other species, the daughter cells are held together by a mucilaginous matrix. As a result, *unbranched filaments* or *amorphous colonies* of diatoms may be formed, and equally simple multicellular thalli occur among green and blue-green algae, although these forms are less common in the sea than in freshwater habitats. Green algae also form more elaborate and complex colonies with a precise form and cell number (the *coenobia* of *Volvox*, *Pediastrum* and *Hydrodictyon*), but this type of thallus is not found in any marine species. Colonies or simple unbranched filaments may be either free-floating in the plankton or attached to a substrate, but the more complex types of thallus are seen only in attached, benthic plants. Irregularly *branched filaments* with an axis formed of single cells joined end to end (*uniseriate*) are found among green, red and blue-green algae, but even the simplest brown alga shows a differentiation of such a branched filamentous thallus into distinct prostrate filaments, which attach the plant to its substrate, and erect filaments, which grow away from the substrate. The relative development of the prostrate and erect systems in such a *heterotrichous* thallus is, however, often uneven; it may be difficult to identify the prostrate system in some species (e.g. the brown alga *Ectocarpus*), whereas in others the erect system may be completely lacking. The *crustose* thalli, that are frequently found on rocks in the intertidal zone, are formed by the fusion of the prostrate filaments in a heterotrichous thallus to produce a 2-dimensional plate of cells closely appressed to the substrate. Recent culture studies have shown that many of these crustose species are stages in the life history of an erect plant (e.g. *Ralfsia/ Scytosiphon*; *Petrocelis/Gigartina*; see Chapter 5). This fusion and interweaving of filaments can also occur in erect systems to produce a coherent **pseudo-parenchymatous** thallus. This morphology is typical of most of the larger red algae, and is also found in some brown algae and in some siphonaceous green algae (see below), but the only other plants with such morphology are the fruiting bodies of higher fungi (i.e. 'toadstools'). These fungi are terrestrial, but pseudo-parenchymatous thalli are almost never found in photosynthetic plants outside the marine flora.

In truly **parenchymatous** thalli, cell division may occur in any plane, and a coherent 2- or 3-dimensional tissue is built up in this way. Parenchymatous sheets of cells, one or two cells thick, occur in green, red and brown algae, but thicker, 3-dimensional parenchymatous thalli are found only in brown algae, and all the very large seaweeds (kelps, wracks, rockweeds, etc.) have this morphology. Among the algae, therefore, this type of thallus is also confined to the sea. The largest and most complex green algae are **coenocytic** or siphonaceous forms, in which the thallus consists of tubular filaments that are not divided into cells. The simplest types are short, unbranched filaments or sacs, but larger thalli may be built up by the prolific branching of a single filament (**uniaxial**), or by weaving branched filaments together into a **multiaxial** pseudo-parenchymatous thallus. Again, almost all of these siphonaceous species are marine, so that this third type of macroscopic thallus construction is also largely confined to the sea.

Thus, the evolution of algal morphology has progressed considerably further in the sea than in freshwater or sub-aerial habitats. The most advanced form of thallus to be found in non-marine algae (apart from the Charophyta, which perhaps should not be considered as algae[19]) is the heterotrichous filamentous construction of the Chaetophorales (Chlorophyta). This contrast between algal morphology in marine and freshwater habitats raises another fundamental question for the marine botanist. Why are the more complex algal thalli found only among the seaweeds, whereas the more complex morphologies among other plant groups (vascular plants, Basidiomycetes) are confined to the land or to fresh water?

ECOLOGICAL CLASSIFICATION

The subject of this chapter — and indeed this whole book — is based on the idea that plants can be classified as 'marine' or 'non-marine' according to their ecological habitat. The boundaries between biological categories are never clear-cut, and the distinction between 'marine' and 'non-marine' plants is bound to be somewhat arbitrary. The plants which are considered to be marine in this book are those that have an absolute requirement for regular or continuous immersion in sea water. This definition excludes the many species of flowering plants which show a **maritime** distribution, or which can tolerate (but do not require) high concentrations of salt. The biology of such **halophytes** has been fully discussed in recent books by Waisel[256] and Reimold and Queen.[209] Our definition also excludes those algae which are characteristic of **brackish water** habitats (i.e. those with salt concentrations intermediate between fresh water and sea water), unless these species are also found in fully marine environments. Since very few species of plants, and relatively few of the larger taxa (see p. 2; Table 1.1), are able to grow in both freshwater or terrestrial habitats and under truly marine conditions, this definition appears to provide the most natural boundary between 'marine' and 'non-marine' plants.

Marine plants, then, are plants that characteristically grow in sea water but, since the vast majority of these plants are photosynthetic, they cannot grow *wherever* there is sea water. Photosynthetic marine plants are necessarily restricted to illuminated sea water, or the surface layer of the sea which is known as the **photic zone.** The depth of this zone depends on the degree of light penetration through the water; this is discussed in detail in Chapter 2, but an average depth for the photic zone is about 100 m. Since the mean depth of the world's oceans is close to 4000 m, photosynthetic plants are restricted to the upper 2.5% of the mean depth, and the majority of the primary productivity in the seas occurs in an even shallower surface layer (see Chapter 4). Heterotrophic plants, and especially fungi and bacteria, are not restricted to the photic zone, and they undoubtedly play an important ecological role as decomposers in sediments at all depths (see Chapter 9), but this book is primarily concerned with the illuminated one-fortieth of the seas.

These seas can be divided into two broad regions with contrasting physical conditions. About 10% of the total area of the sea overlies the shelves surrounding the major continents, and is relatively shallow (less than about 200 m). This region is described as *'coastal'* or 'neritic', whereas the deeper water beyond the continental shelves is described as *'oceanic'* (Fig. 1.1). Coastal waters receive run-off from the adjacent land-masses

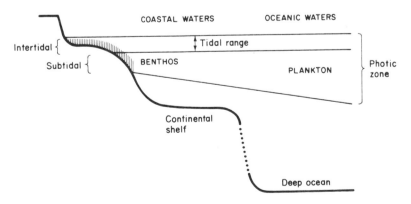

Fig. 1.1 The habitats occupied by marine plants (not to scale).

and, therefore, contain higher concentrations of inorganic nutrients than oceanic waters. This stimulates primary production, and the increased organic matter, combined with inorganic particles stirred up from shallow sediments by wave action, greatly increase the turbidity of the water. Thus, light penetration is reduced and photosynthesis may become impossible only 10 m below the surface. In oceanic waters, on the other hand, photosynthetic plants have been found growing at depths of nearly 200 m. The contrast between coastal and oceanic waters is most extreme in the vicinity of estuaries, where inorganic nutrient concentrations and turbidity tend to be highest, and reductions in salinity frequently add another dimension to the ecological variety.

Within each of these regions, two basic habitats can be recognized: *planktonic* and *benthic* (Fig. 1.1). The phytoplankton consists entirely of microscopic, free-floating plants (unicells, colonies or filaments) at the mercy, so far as their horizontal distribution is concerned, of the movement of the water in which they are suspended. Those organisms which are motile, such as the dinoflagellates and coccolithophorids, and those which can adjust the buoyancy of their cells (e.g. blue-green algae and some diatoms), may exert some control over their vertical distribution, and different species may thus become concentrated at different depths, but they can have little control over their horizontal distribution. Benthic plants are attached or sedentary forms ranging in morphology from non-flagellated unicells (mainly diatoms) through every morphological type listed in Table 1.3 to the massive parenchymatous seaweeds. Benthic habitats are frequently subdivided according to two criteria: the degree of exposure to the air that the plant receives, and the substrate to which the plant is attached. Habitats that are never completely submerged by the tides are described as *supratidal*, while those that are never exposed to the air are *subtidal*. The region in between, all of which is exposed to the air at some times and is submerged at others, is referred to as the *intertidal* (Fig. 1.1). The biological classification of the zonation patterns of rocky shores, mostly based on the term 'littoral', is discussed in Chapter 6, together with the physiology and ecology of intertidal plants.

The classification of benthic plants according to substrate utilizes a series of specific terms, which are largely self-explanatory but are, nevertheless, often misused. They make use of the prefixes *epi-* (on, upon) or *endo-* (inside, internal), and the terms most commonly encountered are shown in the table below.

Substrate	Growing on surface	Growing within
Mud, fine sediments	Epipelic	—
Sand grains	Epipsammic	—
Rocks	Epilithic	Endolithic
Plants	Epiphytic	Endophytic
Animals	Epizoic	Endozoic

Epipelic and *epipsammic* organisms are necessarily microscopic forms, and the most important algae in these habitats are benthic diatoms.

Epipsammic diatoms tend to be small (less than 20 μm), and show only limited motility. They accumulate in the cracks and depressions of sand grains, where they are less likely to be crushed during sand movement, and they may achieve densities as high as 2600 cells mm $^{-2}$ of grain surface. The diatoms associated with finer sediments are larger than epipsammic forms and cannot attach themselves to individual particles. They are also more motile, and migrate actively to the mud surface when it is exposed during daylight, and descend again just before the tide comes in.

Macroscopic algae require a more secure attachment than that offered by sand or mud, and most seaweeds are either attached directly to rocks or boulders (**epilithic**) or to other organisms which are themselves epilithic. **Endolithic** algae dissolve away the surface of rocks or the calcareous deposits of other organisms (e.g. coral, shells) and so come to live within their hard substrate. Such algae are often described (rather ambiguously!) as boring algae, and the commonest examples are the green filaments of *Ostreobium*, which are regularly found beneath the living cells of corals, and the filamentous conchocelis-phase of the red alga *Porphyra*, which, as its name suggests, occurs mainly in shells.

Many of the plants which grow attached to other plants (**epiphytic**) or to animals (**epizoic**) have no more than a superficial attachment, and they are generally assumed to be completely independent of their host organism. Some epiphytic algae, however, are always associated with a single host species, and this specificity suggests that the epiphyte is dependent on the host in some way. Other epiphytes obtain a firmer grip on the host plant by growing down into the host's tissues, and so becoming to some extent **endophytic**. Such penetration of the host has been regarded as evidence of parasitism in some common associations (e.g. the red alga *Polysiphonia lanosa* on the brown seaweed *Ascophyllum*), but it seems unlikely that epiphytes with well-developed photosynthetic tissues obtain a significant proportion of their organic carbon from the host. Some species of endophytic brown and red algae, however, consist of little more than colourless pustules on the outside of the host, and these are undoubtedly **parasitic**. About 80 species of parasitic red algae have been described, and they are unusual among parasites in that they are all closely related to their hosts. Nearly 75% of these species belong to the same taxonomic family as their host (adelphoparasites), and none of them attacks hosts outside the Rhodophyta. Parasitic brown algae (species of *Ectocarpus* and related genera), on the other hand, attack unrelated brown algae or red algae. Certain groups of dinoflagellates are also important as parasites, but their hosts are all marine invertebrate animals — sponges, flatworms and molluscs. Other **endozoic** dinoflagellates retain their chloroplasts and maintain a mutually beneficial symbiotic association with their animal hosts, and coccoid green algae and some diatoms may also be found as endozoic parasites or symbionts.

This account of the taxonomic, morphological and ecological classification of marine plants has been designed to introduce many of the species, and to define many of the terms, that are referred to in later chapters. There is no attempt here, however, to provide a more detailed systematic treatment of particular groups; the reader is referred to Bold and Wynne[19] for an up-to date account of the taxonomy and morphology of the algae. The rest of this book concentrates on those aspects of the physiology and ecology of marine plants which specifically equip them for life in the sea, and which separate them from non-marine plants.

2

The Sea as an Environment for Plant Growth

A plant growing at sea level under clear skies will receive only about 20% less light than a plant growing at the same latitude at an altitude of 4000 m. The same percentage reduction of visible radiation occurs within about 2 m of the clearest oceanic water, and within less than 20 cm of turbid coastal water. Thus, although both air and water are commonly regarded as transparent and colourless media, light penetration through water is at least 2000-fold less than that through air. Light will, therefore, limit plant growth far more often under water than on land, and so it is necessary to examine the behaviour of light in the sea in some detail. This discussion will lead to a definition of the 'photic zone' — the region of the sea to which photosynthetic plant growth is necessarily restricted — and the subsequent treatment of other aspects of the marine environment can largely be limited to this surface layer of the sea.

LIGHT IN THE SEA

Measurement

Light measurement in botanical studies has been bedevilled for many years by the existence of the 'light meter' and a system of 'photometric' standards and units (lumen, lux, foot-candle, etc.). These instruments and units were specifically designed for the spectral response of the human eye, and may give very misleading results if applied in a general biological or ecological context. Strictly, the plant physiologist is not concerned with 'light' — which is radiation containing only those wavelengths that he can see — but with either solar radiation (290 – 3000 nm at the earth's surface) or 'photosynthetically active radiation' (P.A.R.; 350 or 400 to 700 nm). This radiation can be measured either as *irradiance* (i.e. rate of arrival of energy per unit area of surface, units: $J\,m^{-2}s^{-1} = W\,m^{-2}$) or as *photon irradiance* (i.e. rate of arrival of visible quanta per unit area, units: $mol\,m^{-2}\,s^{-1}$, where mol = one mole or Avogadro's number of photons).

As radiation passes through water, it is subject to both absorption and scattering, and the overall reduction in irradiance which results from these two processes is called **attenuation.** The attenuation (or 'extinction') coefficient for a water sample is calculated from two measurements of irradiance at different depths:

$$\text{attenuation coefficient, } k\ (m^{-1}) = \frac{\log_e I_1 - \log_e I_2}{d_2 - d_1}, \qquad (2.1)$$

where I_1, I_2 = irradiance at depths 1 and 2;
d_1, d_2 = depths of two measurements $(d_2 > d_1)$ in metres.

The irradiance at any depth (z, in metres) can then be calculated from:

$$\log_e I_z = \log_e I_o - k.z, \qquad (2.2)$$
where I_o = irradiance at the surface.

Alternatively, the **transmittance** of light through 1 m of water can be measured and used to calculate the attenuation coefficient, thus:

$$\text{transmittance, } T\ (m^{-1}) = \frac{I_{(z+1)}}{I_z}, \text{ and } k(m^{-1}) = -\log_e T.$$

Spectral composition of underwater light

The attenuation coefficient of pure water (Fig. 2.1) is smallest at about 465 nm, and increases towards both ultra-violet and infra-red wavelengths. If the attenuation coefficient for any wavelength is known, it is a simple matter to calculate the depth of water that will absorb a given proportion of the light at the surface, and the right-hand axis of Fig. 2.1 shows the depth to which 0.1% of the surface irradiance at each wavelength will penetrate. Approximately half of the energy content of solar radiation at sea level is in the infra-red region (>700 nm), and the bulk of this radiation is absorbed by 1 m of water. This accounts for the strong heating effect of solar radiation on the surface of the sea, and it also means that, below the top metre of the sea, almost all natural radiation will be in the photosynthetically active range. Absorption and scattering also increase sharply in the ultra-violet region, but most near-UV radiation (310 – 380 nm) will penetrate through several metres of pure water, and a similar depth of clear oceanic water.

Sea salts have very little effect on the optical properties of water and so deep ocean water from unproductive regions (e.g. Sargasso Sea), which is filtered to remove all fine particles, has an attenuance spectrum that is almost the same as that of pure water. The presence of fine particles increases the scattering of the light and, since scattering is inversely proportional to the fourth power of the wavelength, the effect is most marked in the blue and ultra-violet regions of the spectrum. Larger particles, such as microorganisms and biological detritus, will both scatter and absorb light

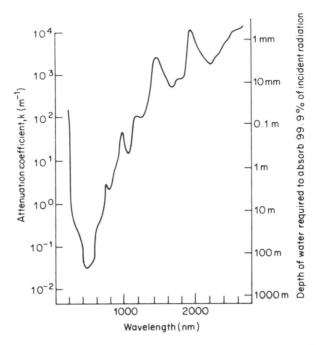

Fig. 2.1 Attenuance spectrum for pure water between 200 and 2800 nm. Right-hand axis shows the depth at which the irradiance is reduced to 0.1% of its surface value. (After Morel, A. (1974). In *Optical Aspects of Oceanography*, Jerlov, N.G. and Steeman Nielsen, E. (eds), pp. 1–24. Reproduced with permission. ©Academic Press inc. (London) Ltd.)

and, although scattering by large particles is independent of wavelength, absorption is strongest in the blue. Thus, more productive oceanic waters show an increased attenuance throughout the spectrum, but there is a greater reduction of blue and ultra-violet wavelengths. Coastal waters tend to be still more productive and to receive run-off from the land; these two factors result in an increase in suspended particles and in the concentration of dissolved organic substances in the sea. This dissolved material is collectively known as 'yellow substance' (or 'gelbstoff') and, as this name suggests, it also absorbs blue light more strongly than other wavelengths. Thus, the decrease in overall transmittance which occurs during the progression from clear oceanic waters to turbid coastal waters is accompanied by a shift of the wavelength at which maximum transmittance occurs, from about 465 nm in clear waters to 575 nm in turbid waters.

Such variations in spectral transmittance have provided the basis for an optical classification of sea water.[107] Oceanic waters are divided into three 'water types', designated by Roman numerals, while coastal waters are divided into nine types, and given Arabic numbers. In each group of water types, the higher the number, the poorer the light penetration and the longer the wavelength of maximum penetration. The curves representing the spectral transmittance per metre in each water type (Fig. 2.2) are a reasonable approximation of the spectral composition of natural light at a depth of 1 m but, as depth increases, these broad curves are rapidly converted into sharply peaked distributions (Fig. 2.3a). At 10 m depth in most coastal waters, about 70% of the total quanta are concentrated within a waveband of

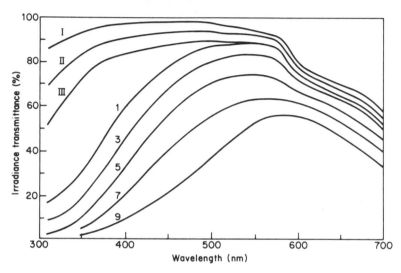

Fig. 2.2 Transmittance per metre of downward irradiance in the surface layer of selected oceanic (I – III) and coastal (1 – 9) water types.[107]

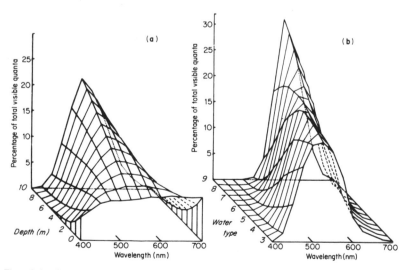

Fig. 2.3 Spectral distribution of underwater irradiance in coastal water types, expressed as the percentage of total visible quanta (400 – 700 nm) in each 25 nm waveband. **(a)** Variation with depth (surface to 10 m) in water type 5; **(b)** variation with water type (types 3 to 9) at 10 m depth[60].

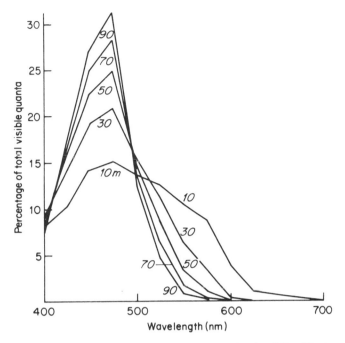

Fig. 2.4 Spectral distribution of underwater irradiance at depths of 10 to 90 m in oceanic water type I.

100 nm bandwidth. The differences between the water types are also intensified with depth (Fig. 2.3b), since the broad peak at 500–550 nm in clearer coastal waters (type 3) shifts to a sharp peak at 575 nm in the most turbid waters (type 9). A similar intensification of the spectral distribution occurs with depth in oceanic waters but, because of the greater transparency of these water types, the distributions remain fairly broad until about 30 m. At this depth, however, less than 0.3% of the total quanta will be at wavelengths greater than 600 nm (Fig. 2.4). Thus, light becomes increasingly monochromatic as it penetrates the sea and oceanic waters can be broadly described as 'blue', whereas coastal waters are 'green'.

There have been few studies of the geographical distribution of different water types, apart from the original world-wide survey on which the classification is based, but some broad generalizations are possible. Oceanic water type I occurs in nutrient-poor, unproductive regions, such as the Sargasso Sea and the eastern Mediterranean, whereas types II and III prevail in more eutrophic waters, and particularly in areas of upwelling.[107] There is less information on where different coastal water types occur, but open coasts seem to have clearer waters than enclosed areas, such as the North Sea and the Baltic. Water type may also change with season. Coastal type 5 occurs during the summer near Helgoland in the North Sea, but autumn and winter storms cause a deterioration to type 7 or even type 9,[146] and a similar winter decline in transmittance has been observed near the Isle of Man.[117]

Intensity of underwater light

The previous section has emphasized the changes in quality that occur as light penetrates sea water of different 'types', but the changes in the overall quantity — or intensity — of the light are probably more significant for plant growth. The transmittance of oceanic water type I reaches a maximum of 98.2% m^{-1} at 465 nm (Fig. 2.2). This corresponds to an attenuation coefficient of 18×10^{-3} m^{-1}, and the depth at which radiation of this wavelength is reduced to 1% of its surface irradiance (the '1% depth') can be calculated from equation 2.2 as 254 m. The maximum transmittance of coastal water type 9 is only 56% m^{-1} at 575 nm, and the 1% depth for this wavelength in this water type is only 7.9 m. These calculations assume that the transmittance of the water for specific wavelengths does not change with depth, and this assumption appears to be justified for well mixed, homogeneous waters. Unfortunately, the same assumption does not hold for the transmittance through water of 'white' light, since this consists of a mixture of wavelengths. The transmittance of P.A.R (350 – 700 nm) increases with depth in all water types. This is not because the spectral properties of the water change with depth, but because the spectral composition of the light is progressively changed as it penetrates the water (see Fig. 2.3a). Since the intensity of the more strongly attenuated wavelengths is reduced more rapidly than that of the other wavelengths, the light becomes progressively richer in the wavelengths that penetrate furthest. Thus, the transmittance of natural light at the surface is equivalent to the mean transmittance of all the constituent wavelengths but, as the light becomes more and more monochromatic, the overall transmittance gradually approaches that of the most penetrating wavelength. This process is illustrated in Fig. 2.5. If transmittance did not change with depth, a plot of the logarithm of irradiance against depth would be linear, since the slope of the graph (i.e. k in equation 2.2) would be constant. In all water types, however, such plots (Fig. 2.5) show a marked increase in slope with depth near the surface, and only gradually approach linearity.

The lower depth limit for large brown algae — mostly members of the Laminariales — in a number of different water types roughly corresponds with the 1% depths determined from Fig. 2.5 (e.g. 95 m in the Mediterranean — type IA; 25 m on the French Atlantic coast — type III; 8 m near Helgoland — type 7[146]), and the compensation depth for phytoplankton occurs at the same level of irradiance. Smaller seaweeds, such as crustose red algae, penetrate more deeply than the Laminariales, and their lower depth limit corresponds in several water types with the 0.05% depth.[146] Thus, the lower limit of the photic zone for phytoplankton and Laminariales ranges from 105 m to 6 m, and that for small benthic multicellular algae ranges from about 170 m to about 10 m (Table 2.1). These depths represent the absolute limits for photosynthetic plant growth in the sea, and the rest of this chapter is concerned with the variability of other aspects of the marine environment within these limits.

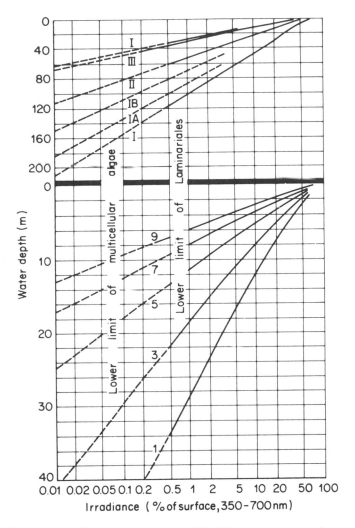

Fig. 2.5 Depth profiles of total irradiance (350 – 700 nm) in a range of water types (oceanic waters above; coastal waters below), showing approximate positions of critical light levels.[146]

TEMPERATURE AND SALINITY WITHIN THE PHOTIC ZONE

These two factors are best considered together because they interact in their effects on the density of sea water, and this property has a primary influence on the stability of a water column and on the movement of water masses within the oceans. In the open oceans, both the temperature and the *salinity* (i.e. the total content of dissolved inorganic salts, usually expressed as parts per thousand — ‰ — or g l^{-1}) of surface water are strongly dependent on latitude (Fig. 2.6). Surface temperatures are greatest near the

Table 2.1 Depth of photic zone for phytoplankton and Laminariales (1% depth) and for multicellular algae (0.05% depth) in different water types (based on data from[107, 146]).

Water type	1% depth (m)	0.05% depth (m)
Oceanic		
I	105	175
II	55	95
III	32	55
Coastal		
1	27	48
3	17.5	31.5
5	11.5	20
7	8.0	14
9	6.0	10.5

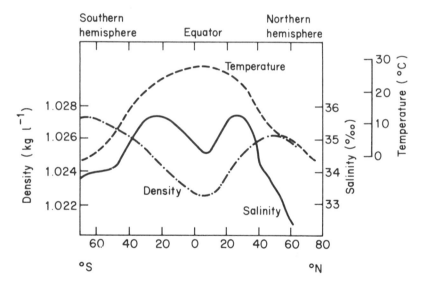

Fig. 2.6 Temperature, salinity and density of surface sea water at different latitudes. (From Pickard, G.L. (1979). *Descriptive Physical Oceanography*, 3rd edn. Pergamon Press, Oxford.)

equator (about 28°C) and decline steadily between the tropics and the poles, reaching a minimum of about −2°C in water containing melting ice. Maximum salinities, on the other hand, are recorded near the tropics, at about 25°N and S of the equator, where the trade winds cause high evaporation rates. Higher precipitation results in a pronounced decrease in salinity, both towards the equator and towards higher latitudes. The total salinity range in the open oceans (33 − 37 ‰), however, is small compared with that for surface temperatures, and the variations in the density of ocean water are mainly related to temperature, as can be seen in Fig. 2.6.

Local variations of both temperature and salinity often occur in coastal waters. Surface temperature may be affected by north – south currents, which flow along the eastern and western margins of most oceans, and by upwelling of cooler subsurface water, which is often observed in localized areas off the western coasts of major land masses. As a result of these influences, surface temperatures in coastal waters are frequently correlated with longitude, rather than with latitude. The salinity of coastal waters may be substantially reduced by the discharge of fresh water from rivers, and these effects will be most pronounced close to major estuaries and at times of high river flow. Under these conditions, large variations in salinity may be associated with fairly uniform temperatures, so that here salinity often exerts a major effect on density. Thus, the water of low salinity flowing out of an estuary frequently forms a distinct layer floating on top of the denser, more saline sea water. Salinities outside the typical oceanic range may also occur in shallow or enclosed seas which are subject either to large inflows of fresh water (e.g. Baltic Sea) or to high evaporation rates (e.g. eastern Mediterranean, 39‰; Red Sea, 41‰).

The ultimate source of heat for the whole of the oceans is, of course, the sun, but the strong absorption of infra-red radiation by water (Fig. 2.1) means that it is only the top few metres which benefit from a direct warming effect. As the temperature of the surface water increases, so its density decreases, and this inhibits the mixing of the warmer surface waters with the cooler, heavier waters beneath. Thus, two layers of reasonably uniform but quite different temperature are formed in the sea: a warm surface layer, mixed by wind action and surface currents; and a cool deep layer with its own quite separate system of circulation (Fig. 2.7). Between these two layers, there is a zone, known as the main **thermocline,** in which the temperature decreases and the density increases rapidly with depth, and the

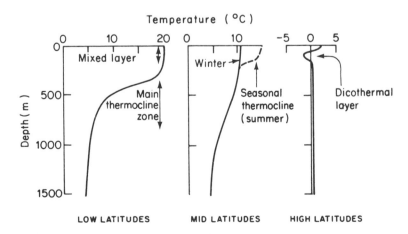

Fig. 2.7 Typical depth profiles of mean temperature in the open ocean. (From Pickard, G.L. (1979). *Descriptive Physical Oceanography*, 3rd edn. Pergamon Press, Oxford.)

steep density gradient that occurs in this zone effectively prevents any exchange of water between the deep ocean and the surface layer. This has important implications for the availability of inorganic nutrients for plant growth (see p. 31).

Since the surface waters at low latitudes are warmer than those in temperate regions, the fall in temperature across the thermocline is much greater in tropical waters (Fig. 2.7), but the seasonal increase in solar radiation that occurs in middle latitudes in the summer causes a secondary, seasonal thermocline to be formed within the layer defined by the main thermocline. This seasonal thermocline begins to develop in spring and gradually moves towards the surface and intensifies through the summer (Fig. 2.8). Strong cooling of the surface waters in the autumn, combined with a seasonal increase in wind strength, results in the sudden collapse of the seasonal thermocline, and a return to fully mixed conditions throughout the surface layer in late autumn. This pattern of growth and decay of a seasonal thermocline is shown in three different ways in Fig. 2.8. Any of these ways may be used to present the variations with depth and time of a particular characteristic or property of the sea, and a detailed comparison of these three graphs will make similar figures on later pages easier to understand and to interpret.

In polar waters, the temperature at the surface is so low that there is little difference between the surface and deep water temperatures, and no thermocline can be detected (Fig. 2.7). The increased solar radiation in the summer is largely dissipated in melting ice, and little change in surface temperature occurs, but the temperature profile may show a subsurface minimum (the dicothermal layer) at about 100 m.

The changes in salinity that occur with depth in the open ocean and in most coastal waters do not show such a clear pattern as the temperature variations, and probably have little effect on plant growth. Since temperature is the main factor controlling density in such waters, there is no salinity equivalent of the thermocline, and salinities within the general oceanic range of $33-37°/oo$ may be found at any depth. It is only in the neighbourhood of large estuaries that biologically significant changes of salinity with depth may be observed.

In spite of the strong absorption of solar radiation by surface waters, their temperatures in the open oceans rarely change through the day by more than $0.3°C$. This is because of thorough mixing within the surface layer, and because much of the absorbed energy is lost through evaporation. In shallower or more sheltered waters, the diurnal temperature range may reach $2-3°C$, but even this is only one-tenth of the range that is commonly experienced close to the ground in terrestrial habitats. The seasonal variations in surface temperatures are equally moderate in comparison with terrestrial conditions. At the equator, the annual temperature range in oceanic waters is a mere $2°C$, and this increases with latitude to $8-9°C$ at $50°N$ or S (Fig. 2.8), and then decreases again towards the poles. The largest annual temperature range for marine habitats is recorded from sheltered coastal waters in temperature climates and amounts to $10-15°C$.

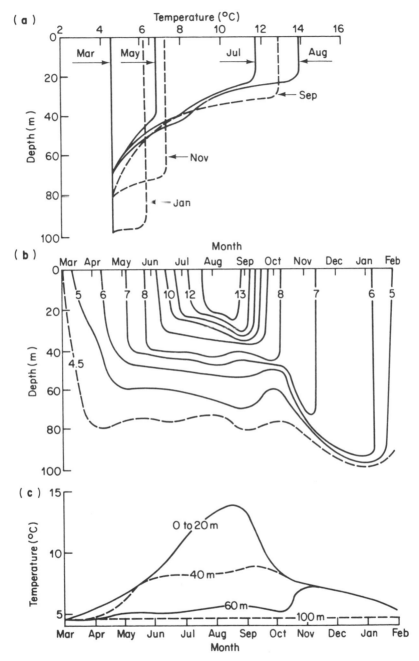

Fig. 2.8 Growth and decay of the seasonal thermocline in the eastern North Pacific (50°N, 145°W). **(a)** Depth profiles of temperature for selected months through the year; **(b)** depth-time diagram, showing temperature contours at 1°C intervals; **(c)** seasonal temperature variation at different depths. (From Knauss, J.A. (1978). *Introduction to Physical Oceanography*. Reproduced by permission of Prentice-Hall, Inc., Englewood Cliffs, N.J.)

These annual and diurnal variations of temperature apply to surface waters, and the ranges decrease rapidly with depth. Water temperature is essentially constant throughout the day at depths below $6-8$ m, and constant throughout the year at depths greater than $100-300$ m. This last figure suggests that the whole of the photic zone is subject to some seasonal variation of temperature, but this is often very small. The sea is a very constant, equable environment compared with the land.

CHEMISTRY OF THE PHOTIC ZONE

Quantitatively, the chemistry of sea water is dominated by a small number of inorganic salts whose collective concentration is described as the salinity. The composition of this mixture of salts is almost the same in waters of different salinity, and so the concentration of any one of them can be estimated from a single determination of their total concentration. Although, in the past, salinity had to be measured by evaporating sea water to dryness and weighing the residue, modern 'salinometers' simply measure the conductivity of sea water and so provide a far more accurate and convenient estimate of salinity.

Eleven elements or their ions make up 99.7% of the total material dissolved in the sea. These elements are known as the 'major constituents'. They all have concentrations greater than about 10^4 molal, and their concentration relative to the salinity of the water ('salinity ratio') does not vary by more than about 10%, except for the least abundant of these elements, fluorine (Table 2.2). This constancy of composition of these major constituents provides evidence that the rate of overall mixing of the oceans is fast relative to the rates at which these elements are being added to or removed from the sea. It also indicates that any biological utilization of these elements is insignificant in comparison with the total amounts available in sea water.

If the total charge of the cations in Table 2.2 (calculated from the product

Table 2.2 Concentrations of major constituents in sea water of 34.32°/$_{oo}$ salinity, and their salinity ratios (adapted from Pytkowicz, R.M., Atlas, E. and Culberson, C.H. (1977). *Oceanography and Marine Biology, Annual Reviews*, **15**, 11 – 45).

Constituent	Concentration mg-at l^{-1}	mg l^{-1}	Salinity ratio mg l^{-1} (°/$_{oo}$)$^{-1}$	Range
Chloride, Cl$^-$	535.0	18971	552.8	—
Sodium, Na$^+$	459.1	10555	307.5	306.1 – 308.7
Magnesium, Mg^{++}	52.86	1268	36.95	36.8 – 37.3
Sulphate, SO$_4^{--}$	27.67	2657	77.42	77.2 – 77.8
Calcium, Ca^{++}	10.07	403.9	11.77	11.7 – 11.9
Potassium, K$^+$	10.0	391	11.39	11.2 – 11.6
Bicarbonate, HCO$_3^-$	2.33	142	4.138	3.91 – 4.38
Bromide, Br$^-$	0.825	65.9	1.920	1.90 – 1.94
Boron, B	0.421	4.48	0.131	0.123 – 0.146
Strontium, Sr^{++}	0.088	7.70	0.224	0.210 – 0.244
Fluoride, F$^-$	0.067	1.3	0.038	0.035 – 0.050

of the molarity and the valency of each) is compared with the total charge of the anions, excluding bicarbonate, it will be seen that there is a slight excess of positive over negative charges. It is this excess of positive charges which creates what is known as the *total alkalinity* of sea water:

$$\text{total alkalinity, } [A] = [Na^+] + [K^+] + 2[Mg^{++}] + 2[Ca^{++}] - [Cl^-] - 2[SO_4^{--}] - [Br^-]. \tag{2.3}$$

Since sea water is electrically neutral, this excess positive charge must be balanced by the variable concentration of negative charges contributed by carbon and boron. Both of these elements can exist as anions with different charges, so that their contribution to the total negative charge can vary while the absolute concentration of the element remains constant. Since the concentration of boron in the sea is substantially lower than that of carbon (Table 2.2) and the major ions of boron become significant only at a higher pH than that of typical sea water, it is the various forms of carbon and the resultant *carbonate alkalinity* which are mainly responsible for neutralizing the excess of positive ions in sea water (see below).

Most of the remaining substances dissolved in sea water (0.3% of the total) are very much more variable in concentration, and this variability can usually be attributed to biological activity. These substances fall into two main categories: dissolved gases and minor constituents.

Dissolved gases: nitrogen, oxygen and carbon dioxide

The solubilities of the major gases of the atmosphere are such that the composition of the total gas dissolved in sea water, in equilibrium with air, is very different from that of the atmosphere (Table 2.3). The $N_2:O_2$ ratio is decreased by about half, and the $CO_2:O_2$ ratio is increased 5-fold. There is almost as much molecular carbon dioxide dissolved in a litre of sea water as there is in a litre of air, but the absolute concentrations of nitrogen and oxygen in sea water are about 1/80th and 1/40th, respectively, of their concentrations in air (Table 2.3). The solubility of all gases decreases with increase in temperature and with increase in salinity, but these variations have little effect on the relative concentrations of different gases in natural sea water.

Table 2.3 Concentrations of major gases in the atmosphere and dissolved in sea water (34.5°/$_{oo}$ salinity) in equilibrium with air at 15°C (calculated from values given by Riley, J.P. and Chester, R. (1971). *Introduction to Marine Chemistry*. Academic Press, London).

Gas	Concentration in air		Concentration in sea water		
	ml l^{-1}	% (by vol.)	ml l^{-1}	mg l^{-1}	% (by vol.)
Nitrogen, N_2	780.88	78.1	10.57	13.21	63.03
Oxygen, O_2	209.49	21.0	5.65	8.07	33.69
Argon, A	9.30	0.9	0.28	0.50	1.67
Carbon dioxide, CO_2	0.320	0.032	0.27	0.53	1.61

Molecular nitrogen is a relatively inert chemical, but it is involved in two opposing microbiological processes — nitrogen fixation and denitrification. The magnitude of both of these processes is difficult to estimate in the sea, but their net effect must be small, since the nitrogen content of sea water rarely departs from saturation by more than 10%. Oxygen is also involved in two biological processes of opposite nature — respiration and photosynthesis — but, since the magnitude of these processes is greater, and they may become spatially separated in the sea, the concentration of oxygen in sea water is far more variable than that of nitrogen. Photosynthesis generally exceeds respiration in surface waters, especially during the day in summer months, and the oxygen content is, therefore, close to or in excess of saturation. Surface mixing ensures a fairly uniform oxygen concentration throughout the water above the thermocline, although a subsurface oxygen maximum is sometimes observed in sheltered waters. At the thermocline, or towards the lower limit of the photic zone, oxygen production through photosynthesis declines, but consumption and decomposition of organic material formed in the surface layers continue. Therefore, oxygen concentration frequently exhibits a sharp decrease at about the same depth as the thermocline, and only slowly increases again at greater depths. Within the photic zone, however, biological activity is unlikely to be limited by a shortage of oxygen.

The same conclusion is often reached about the CO_2-supply in the sea, but the chemical behaviour of carbon dioxide in water is more complicated than that of either oxygen or nitrogen, and it is necessary to consider some of the details in order to understand why sea water provides such a constant CO_2-environment. This will also contribute to an understanding of the conditions under which the CO_2-supply may become limiting in marine habitats, and the significance of certain aspects of the physiology of photosynthesis in marine plants.

The availability of carbon dioxide in water is not simply determined by its solubility because CO_2 reacts with water as well as dissolving in it:

$$CO_2 + H_2O \rightleftharpoons H_2CO_3 \rightleftharpoons H^+ + HCO_3^- \rightleftharpoons CO_3^{--} + H^+ \qquad (2.4)$$

The concentrations of the different components of the CO_2-carbonate system at equilibrium are determined by the dissociation constants for each of this series of reversible reactions. These constants vary with temperature and salinity, but tables of first and second 'apparent' dissociation constants of carbonic acid in sea water[233] enable the equilibrium concentrations of all of the components to be calculated from *any two* of the following measurable properties:

(1) pH of the sea water;
(2) carbonate alkalinity (i.e. $[HCO_3^-] + 2[CO_3^{--}]$) — see below;
(3) total carbon dioxide (ΣCO_2 — i.e. $[CO_2] + [H_2CO_3] + [HCO_3^-] + [CO_3^{--}]$) — estimated by measuring the total amount of CO_2 that is driven off when sea water is acidified;

(4) partial pressure of CO_2 in air which is in equilibrium with sea water (P_{CO_2}).

The total alkalinity of sea water (see p. 25) can be determined by potentio-metric titration, and is more or less constant in relation to salinity. The salinity ratio for alkalinity (the 'specific alkalinity') is about 0.070 in most ocean waters. Thus, for water with a salinity of 34.5‰, the total alkalinity = 0.070 × 34.5 = 2.415 meq l^{-1}. In uncontaminated sea water, the only ions, other than carbonates, which contribute significantly to this total alkalinity are those of boric acid, but their contribution amounts to only between 1 and 6% in the normal pH range of sea water. For example, at pH 8.2 and 15°C, the borate alkalinity would be about 0.09 meq l^{-1}, leaving 2.325 meq l^{-1} as the carbonate alkalinity. If sea water with this alkalinity is equilibrated with air at 15°C, it will acquire a P_{CO_2} of about 320 ppm (Table 2.3). All the other components of the CO_2-system can now be calculated from these two values. The pH will be 8.24, and the total carbon dioxide (ΣCO_2) will be about 2.14 mmol l^{-1}, of which 0.58% will be present as dissolved CO_2 (plus a negligible amount of undissociated H_2CO_3), 90.1% as HCO_3^- ions, and the remaining 9.3% as CO_3^{--} ions (Table 2.4 — 'initial values').

If photosynthesis occurs in this water and some of the dissolved CO_2 is removed, HCO_3^- ions will react with free H^+ ions to replace it:

$$HCO_3^- + H^+ \rightarrow CO_2 + H_2O. \qquad (2.5)$$

This reaction will reduce the H^+ ion concentration, so that the pH will rise, but this effect will be offset by the simultaneous dissociation of other HCO_3^- ions, which liberate sufficient H^+ ions to balance the loss in the first reaction:

$$HCO_3^- \rightarrow H^+ + CO_3^{--}. \qquad (2.6)$$

Table 2.4 CO_2 – carbonate system in sea water (34.5‰ salinity) in equilibrium with air at 15°C, and the effects of adding and removing 0.1 mmoles of molecular CO_2 through biological activity. Carbonate alkalinity = 2.325 meq l^{-1} throughout. All concentrations are in mmoles l^{-1}.

Components	CO_2 Removal (Photosynthesis) New values	$\triangle$ C	Equilibrium Initial values	$\triangle$ C	CO_2 Addition (Respiration) New values
Total CO_2 (ΣCO_2)	2.039	− 0.100	2.139	+ 0.100	2.239
[$CO_2 + H_2CO_3$]	0.0068 (0.33%)	− 0.0056	0.0124 (0.58%)	+ 0.0136	0.0260 (1.16%)
[HCO_3^-]	1.7394 (85.31%)	− 0.1888	1.9282 (90.14%)	+ 0.1728	2.1010 (93.84%)
[CO_3^{--}]	0.2928 (14.36%)	+ 0.0944	0.1984 (9.28%)	− 0.0864	0.1120 (5.00%)
pH	8.46		8.24		7.96

Thus, the removal of 0.1 mmoles of CO_2 from our water sample will produce a fall of only 0.0056 mmoles for the dissolved CO_2 (Table 2.4, left). The difference of 0.0944 mmoles is made up from the bicarbonate pool but, for every ion that is converted into CO_2, another is converted into carbonate. Therefore, the CO_3^{--} ion concentration increases by 0.0944 mmoles, and the bicarbonate concentration falls by double this amount. Since carbonate ions carry two negative charges and bicarbonate ions only one, the total negative charge of the system (i.e. the carbonate alkalinity) remains constant, so that the excess positive charge of the major cations continues to be neutralized (see p. 25). The CO_2 fixed in photosynthesis is not entirely replaced from the bicarbonate pool, however, because the reduction in the overall concentration of the dissolved inorganic carbon results in a shift in the balance of the components of the system and a slight rise in pH (Table 2.4). In spite of this, neither the change in pH nor the fall in CO_2 concentration is as severe as they would have been if all of the carbon for photosynthesis had been derived from the pool of dissolved CO_2 in the water. Similar changes occur in the reverse direction when respiration increases the CO_2-content of the water (Table 2.4, right), and the CO_2 – carbonate system thus acts as a buffer for both dissolved CO_2 and for pH.

It is possible to estimate from Fig. 2.9 the effect of a given amount of photosynthesis or respiration on the pH and on the proportion of the different components of the CO_2 – carbonate system in a water sample which is cut off from the air. The slope of the curve of ΣCO_2 vs pH ('A,') provides a measure of the buffering capacity of the system. Large changes of total CO_2 content are required to effect small changes in pH outside the normal pH range of sea water but, within the range 7.5 – 8.5, the system is more sensitive to CO_2 changes. This sensitivity is substantially increased by a large reduction in carbonate alkalinity, such as may occur in waters of low salinity (e.g. curve for 16‰ – Fig. 2.9, A_2). A change in ΣCO_2 of 0.1 mmoles may change the pH by as much as 0.6 units, compared with only 0.3 units at the higher salinity.

Plants which are dependent on CO_2 as a carbon-source for photosynthesis (i.e. cannot utilize bicarbonate ions) will be able to photosynthesize at CO_2 concentrations down to their CO_2 – compensation point, where the CO_2 fixed in photosynthesis is exactly balanced by that evolved in respiration. The CO_2 concentration at which compensation occurs has rarely been determined for marine plants (see p. 65), but most C-3 plants on land have CO_2 – compensation points at 70 – 100 ppm by volume. One hundred ppm is equivalent to about 5 μmol l^{-1}, and this concentration will occur in normal sea water (i.e. with a salinity of about 35‰ at 5 – 25°C) at pH 8.5 – 8.6 (Fig. 2.9, curve D_1). The ΣCO_2 of sea water at this pH will be about 0.2 mmol l^{-1} lower than that of water in equilibrium with the air (pH 8.25; Fig. 2.9, curve A_1), so that marine plants will be able to assimilate this amount of carbon in photosynthesis without replenishment of the CO_2 from the air, or from water at a lower pH. Since the maximum rates of photosynthesis recorded for marine phytoplankton are about 200 mg C m^{-3} day^{-1}

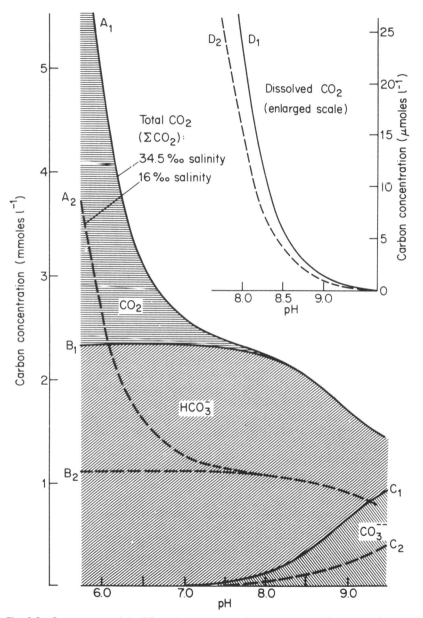

Fig. 2.9 Components of the CO_2-carbonate system in sea water at 15°C as a function of pH. **A**: total CO_2 (ΣCO_2; $[CO_3^{--}] + [HCO_3^-]) + [CO_2]$); **B**: carbonate + bicarbonate ($[CO_3^{--}] + [HCO_3^-]$); **C**: carbonate only ($[CO_3^{-1}]$); **D**: dissolved CO_2 only ($[CO_2]$). The carbonate alkalinity (i.e. $2[CO_3^{--}] + [HCO_3^-]$) remains constant at all pH's, and curves are shown for two values of this property of sea water: 2.325 meq l^{-1}, corresponding to a salinity of 34.5°/oo (curves A_1 to D_1), and 1.11 meq l^{-1} (16°/oo, A_2 to D_2). Inset: concentration of dissolved CO_2 (enlarged scale) at high pH.

(i.e. $16-20$ μmol l^{-1} day^{-1}), it is clear that CO_2-limitation is unlikely to occur in normal sea water. This conclusion is supported by the observation that the pH of open ocean water is rarely outside the range $7.8-8.4$. In waters of lower alkalinity, however, the CO_2-buffering is less efficient, and limiting CO_2 concentrations may be reached after much less photosynthetic activity than in the open sea (e.g. Fig. 2.9, curve D_2).

If plants are able to utilize bicarbonate as a carbon-source for photosynthesis, they will be able to photosynthesize in more alkaline waters, which contain almost no dissolved CO_2. There is evidence that a few marine plants, including benthic algae and phytoplankton, can utilize bicarbonate for photosynthesis (see p. 63), but this ability seems to carry only a minor ecological advantage for plants growing at normal salinities, although it might be more advantageous in estuarine plants. The rarity of the C-4 photosynthetic pathway among marine plants (see p. 65) is also relevant here. A major feature of this type of metabolism in land plants is the reduction in CO_2-compensation point, but marine plants would gain little from C-4 photosynthesis, because there is so much more CO_2 available in the sea than in the air.

Minor constituents: combined nitrogen, phosphorus and silicon

The mean concentrations of selected minor constituents in sea water are shown in Table 2.5, together with the typical range of concentrations for each. The most abundant of these elements, silicon, reaches concentrations as high as some of the major constituents (see Table 2.2), but silicon is far more variable than any of these more conservative elements. Silicon is a major component of diatom cell walls, and the frequent reduction of this element, together with nitrogen and phosphorus, to extremely low concentrations in the sea is caused by the uptake of these elements by plants during growth.

The biological importance of these elements in the sea is perhaps best appreciated by comparing the atomic ratios of nitrogen and phosphorus to

Table 2.5 Concentrations of selected minor constituents in sea water (source as for Table 2.2).

Element	Mean concentration (μg l^{-1})	Typical range of concentration (μg l^{-1})
Silicon	2000	$0-4900$
Nitrogen (combined)	280	$0-560$
Phosphorus	30	$0-90$
Aluminium	6	$0-10$
Iron	3.1	$0.1-62$
Zinc	7.3	$1-48.4$
Iodine	53	$48-80$
Copper	1.3	$0.5-27$
Manganese	1.2	$0.2-8.6$
Cobalt	0.15	$0.005-4.1$

carbon in typical plant material with the ratios of these elements in the sea. An accepted average for the atomic C:N:P ratio in phytoplankton cells is 106:16:1,[236] and the weight ratios derived from these figures are shown in Table 2.6, together with comparable ratios based on typical concentrations of inorganic carbon, nitrogen and phosphorus in sea water. It is clear that the C:N and C:P ratios in the sea are vastly greater than those in phytoplankton cells, and this suggests that the supply of nitrogen and phosphorus is more likely to restrict plant growth in the sea than is the supply of inorganic carbon. This reinforces the similar conclusion reached after examining the CO_2-system in sea water (see p. 30). N:P ratios in the sea are more constant than C:N or C:P ratios, and they are generally similar to the N:P ratios in phytoplankton cells. This observation suggests that nitrogen and phosphorus are taken up from sea water in the proportions in which they occur in plant cells, and that the subsequent decomposition of organic material releases these elements back into the sea in approximately the same proportions. Silicon is not an essential element for all marine plants, but is required only by those, such as diatoms and silicoflagellates, which have silicified cell walls or skeletons. The magnitude of the Si:N or Si:P ratios, either in the phytoplankton or in sea water, is more variable than the N:P ratio, therefore, and largely depends on how many silicon-requiring species are present in the current phytoplankton crop.

The vertical and seasonal distributions of nitrogen, phosphorus and silicon in the sea reflect their biological activity. When the water column is well mixed, and the processes of photosynthesis and decomposition occur side by side, there is little variation of concentration with depth. Whenever a seasonal thermocline develops, however, photosynthesis and decomposition become spatially separated. At these times, the uptake of nutrients is restricted to the surface layers, whereas nutrient release can occur throughout the water column. The steep temperature gradient which marks the thermocline is correlated with an equally steep gradient in the density of the water (the *pycnocline*), and this effectively prevents any mixing between the surface waters and the waters beneath the thermocline. Thus the nutrients removed by plant growth in the surface layers cannot be replenished by decomposition at or below the thermocline, and the concentrations of plant nutrients in the surface waters are rapidly reduced to very low

Table 2.6 Concentration ratios (by weight) of inorganic carbon, nitrogen and phosphorus in phytoplankton and in sea water.

	C:N	C:P	N:P
In phytoplankton cells*	5.7:1	41:1	7.2:1
In sea water:†			
mean concentrations	100:1	932:1	9.3:1
maximal N and P concentrations	50:1	311:1	6.2:1

*Based on atomic ratios for C:N:P of 106:16:1[236]
†Based on values in Tables 2.2 and 2.5

levels, which persist until the thermocline decays in the autumn.

These seasonal variations in nutrient concentrations are illustrated by a complete year's records for phosphate and silicate at a single station in the English Channel (Fig. 2.10). The concentrations of both nutrients are negatively correlated with temperature and, in the summer months, a 'phosphocline' and a 'silicocline' can be identified at the same depth as the

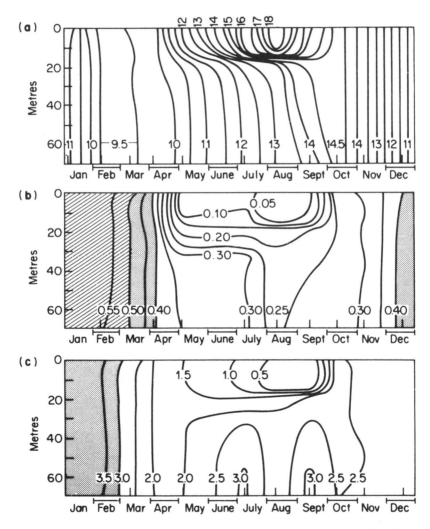

Fig. 2.10 Depth-time diagrams for (a) temperature; (b)phosphate; and (c) silicate concentrations at a single station in the English Channel for 1960. Intervals between contour lines: (a) 0.5°C; (b) 0.05 μg-at P l^{-1}; (c) 0.5 μg-at Si l^{-1}. (From Armstrong, F.A.J. and Butler, E.I. (1962). *Journal of the Marine Biological Association of the United Kingdom*, **42**, 253–8. Reproduced by permission of Cambridge University Press, Cambridge.)

thermocline. Nitrate concentrations show a similar pattern. This nutrient is frequently below the limits of analytical detection in surface waters during the summer, and can be measured only at and below the thermocline. On the other hand, oxygen levels tend to be directly correlated with temperature, and show a marked decrease where the other inorganic chemicals increase (see p. 26).

The solar heating of the surface layer of the sea, which results in the formation of the thermocline and sets up such a strong density barrier between the surface and deeper waters, clearly exerts a fundamental influence on the physics, the chemistry and the biology of the oceans. This influence is made most apparent, perhaps, in areas where it is absent. Upwelling of subsurface waters occurs regularly in localized areas along the eastern shores of tropical oceans, where the combined effect of winds and the earth's rotation causes the surface water to move away from the coast. This water is continually replaced by cooler, nutrient-rich water from depths of $50 - 300$ m, and neither a thermocline nor a nutrient-depleted surface layer is normally established. The increased nutrient concentrations in the surface layers more than offset the decrease in temperature, and the biological productivity in these areas of upwelling may be up to ten times greater than in the warmer, but more strongly stratified, waters of the open ocean (see p. 91).

This discussion of the physics and chemistry of the photic zone of the seas has shown that, in some respects, the sea provides a remarkably constant environment for plant growth. Plants which are growing in the sea will clearly never be short of water, and the supplies of CO_2 and O_2 are also unlikely to be limiting. The temperature of the surface layers of the sea is more variable than deep-sea temperatures, but is very much less variable than the temperature in most terrestrial habitats. Both diurnal and seasonal variations are buffered by the volume and the high specific heat of the water, and organisms have sufficient time to adjust their metabolism to the changes that do occur. The factors that are most variable are irradiance and the supply of inorganic nutrients, such as nitrogen, phosphorus and silicon, and many of the most interesting aspects of the physiology and ecology of marine plants are related to their light and nutrient-harvesting strategies (see Chapters 3 and 4).

These conclusions apply to the plants of the open sea and to subtidal benthic plants, but intertidal organisms are subject to a much less constant environment. Intertidal habitats combine the disadvantages of terrestrial environments — principally the variable temperature regime and unreliable water supply — with the disadvantages of aquatic environments — variability of irradiance and nutrient supply — and are also subject to physical damage caused by wave action. There is a case, indeed, for regarding the intertidal as the most variable environment to be found on earth.

ENVIRONMENTAL PROBLEMS UNIQUE TO INTERTIDAL HABITATS

The diagnostic feature of intertidal habitats is the regular alternation of emersion and submersion that accompanies the ebb and flow of the tides. The tidal movement of the seas throughout the world is caused by the gravitational pull of the moon and the sun, but the pattern of the tides shows substantial variations in different oceans. Along most Atlantic coasts, there are two high tides a day of approximately the same height (Fig. 2.11a) and this regime is described as *semidiurnal*. In a few, mainly tropical localities, however, there is only one tide per day (*diurnal* tides, Fig. 2.11d), and most of the Pacific Ocean experiences so-called *mixed* tides with a varying balance between the diurnal and semidiurnal patterns. Most commonly, this results in two high tides a day, but the water rises very much higher on

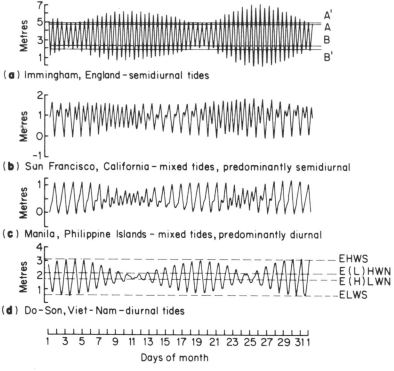

(**a**) Immingham, England – semidiurnal tides

(**b**) San Francisco, California – mixed tides, predominantly semidiurnal

(**c**) Manila, Philippine Islands – mixed tides, predominantly diurnal

(**d**) Do-Son, Viet-Nam – diurnal tides

Days of month

Fig. 2.11 Tidal curves predicted for March 1936 in four sites with different types of tidal regime. In (**a**), lines A/A′ and B/B′ illustrate that points close together on a shore may experience markedly different environmental stresses: A and B are submersed and emersed in every tidal cycle, whereas A′ and B′ may be continually emersed or submersed, respectively, for 3 – 4 tidal cycles. In (**d**), the critical tide levels marked represent the extremes for the month, and are not necessarily the most extreme tides of the year. (Adapted from Dietrich, G. (1963). *General Oceanography: an Introduction*. Wiley-Interscience, New York.)

one tide than on the other (Fig. 2.11b). The height of the water at high tide, and the difference in level between high and low water (the **tidal range**), also vary in any one location on a two-weekly cycle. When the moon is either full or new, its gravitational pull is aligned with that of the sun, and the high tides are higher and the low tides are lower than when the moon and the sun are pulling at right angles to one another, during the first and last quarters of the moon. The most extreme tides in each fortnight are known as **spring** tides (the reference is to spring as a movement, and not as a season) and the least extreme tides are the **neap** tides. Figure 2.11 shows that the same two-weekly cycle of large spring tides and smaller neap tides is superimposed on the daily cycle in all of the different types of tidal regime.

Thermal and desiccation stress

When the tide falls below a particular point on the shore, any plant occupy-ing that point will be transferred suddenly from an aquatic to a terrestrial environment. The most obvious result of this change will be that the plant begins to dry out, but there will also be a rather abrupt change in tempera-ture, and this may exert a more immediate influence. On a hot sunny day, even in cool temperate climates, exposed rock surfaces quickly reach temperatures of $25-30°C$, even though the air temperature may be only around $20°C$ and the water temperature $15°C$ or less. Figure 2.12 shows the temperature record for three heights on a single shore at about $54°N$ over a 24-hour period in late spring. Although the water temperature remained between 10 and $11.5°C$ throughout the day (see p. 22) and the air tempera-ture did not exceed $15°C$, the temperature at the high water mark, close to the rock surface, reached a maximum of $24°C$ and remained above $20°C$ for nearly 4 hours. The temperatures recorded in shaded positions under rocks or seaweeds also climbed steadily during the period of emersion, although they were lower than those in exposed positions. Emersion during the night resulted in a temperature fall of $3-4°C$, compared with the sea water. Thus, during this 24-hour period, plants above mean tide level (MTL) experi-enced a temperature range of $15-17°C$, whereas the range at the low water mark was $6°C$ and that for subtidal plants only $1.5°C$. Although there was little difference in the extreme temperatures recorded at MTL and at the high water mark, plants at the higher level were exposed to these extremes for longer periods (e.g. temperatures of over $20°C$ were recorded for 2 hours longer at the high water mark than at MTL, Fig. 2.12). These tempera-ture changes associated with emersion and submersion will have imme-diate effects on the metabolic rates of intertidal plants, as well as indirect effects on the rates of drying and the degree of desiccation that the plants experience.

Clearly, any plant that is liable to be exposed by the tide for an hour or more in the middle of a summer's day must be able to tolerate a temperature of $30°C$, and a temperature fall of $15-20°C$ within minutes when the tide comes in; therefore, the lowest level exposed by the tide on any shore (the *E*xtreme *L*ow *W*ater of *S*pring tides, or ELWS) must mark a critical

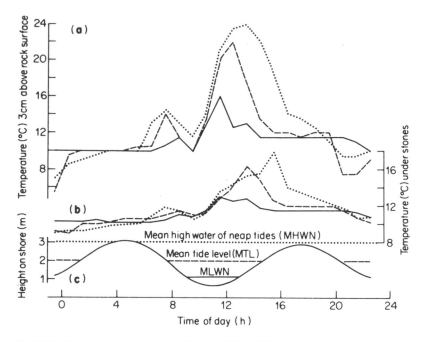

Fig. 2.12 Diurnal temperature variations on 27 May 1981 at three heights on a shore in Belfast Lough, Northern Ireland (unpublished data of J. Brook). One temperature probe at each height (a) was exposed to the air about 3 cm above the rock surface, and a second (b) was located under a stone at the same height. The predicted tidal curve for the same day (c) shows the approximate times of submersion and emersion at each height.

environmental boundary. Above this level, the duration of emersion in each tidal cycle, and the severity of the resulting temperature and desiccation stress, will increase gradually with height up the shore (Fig. 2.13) and it is difficult to define another, equally critical boundary on this basis, apart from the highest point ever submerged by the tides — EHWS. By definition, these two levels, EHWS and ELWS, are submerged and emerged, respectively, by only a few tides in each year, and the level of EHWS, for example, will be continually exposed for the rest of the year. The least extreme tides of the year range from the highest low water of neap tides (E(H)LWN) to the lowest high water (E(L)HWN), and these two levels also mark critical environmental boundaries. This is because there will be at least one tide in the year which will not cover plants living above E(L)HWN, and will not uncover plants below E(H)LWN. Thus, all plants above E(L)HWN will be exposed to the air throughout at least one entire tidal cycle, and plants below E(H)LWN will be continuously submerged for the same period (compare lines A and A′, and B and B′ in Fig. 2.11). If the maximum length of such periods of continuous exposure to the air, or submersion by the sea, is plotted against height on the shore (Fig. 2.13), sharp changes are observed within very short vertical distances, and the ability to withstand these long periods of emersion or submersion may be important in determining the levels that different species can occupy on the shore.

The temperature and drying regime at a particular height on the shore is not determined solely by the predicted tidal regime for that height, and biological boundaries will not correspond exactly to critical tidal levels, even if the major factors governing the distribution of the organisms are thermal and desiccation stress (see p. 123). The exact height at which such stresses begin to operate will be increased on shores subject to wave action, since the splash from breaking waves will clearly moderate the adverse effects of emersion. Similarly, the aspect and the slope of the rock surfaces will affect the intensity of insolation during emersion. A south-facing slope of 30 – 45° will heat up and dry out very rapidly, whereas a north-facing overhang on the same shore could remain cool and moist under exactly the same weather conditions. Finally, the severity of the effects of emersion on the lower part of the shore — just above ELWS — will be significantly affected by the time of day at which the low waters of spring tides occur. Since the tides get approximately 50 minutes later each day, and the spring/neap cycle lasts for 14 days, the lowest tides of one spring series occur about 12 hours later than

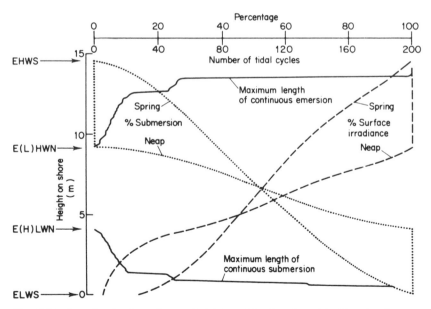

Fig. 2.13 Length of submersion period (dotted lines) and total incident irradiance (broken lines) during the most extreme tidal cycles (spring and neap), compared with the maximum length of periods of continuous emersion or submersion (continuous lines) at different heights on a shore. Submersion is expressed as a percentage of the length of a full tidal cycle, and irradiance as a percentage of the total irradiance (350 – 700 nm) at the surface during a complete cycle, assuming coastal water type 7. The calculations were based on the tidal predictions for 1981 at Avonmouth, which has the greatest tidal range in Europe, but the patterns are similar for all shores with semidiurnal tides.

those of the previous series. In any one locality, therefore, the low waters of spring tides always occur at roughly the same time of day. If this is at about midday or early afternoon (e.g. west coast of Ireland, Plymouth [135]), subtidal organisms will be unable to survive even a short period of emersion but, if low water springs occur at about 6 a.m. and 6 p.m. (e.g. most Irish Sea coasts), the effects of emersion will be far less severe.

Variations in salinity, CO_2, nutrients and irradiance

It is not only the organisms that begin to dry up when the tide goes out. Small bodies of sea water will start to evaporate in hot weather, with the result that salinity may increase substantially. On the other hand, intertidal plants may be exposed either to rain or to streams running across the beach at low tide, so that salinity may also decrease substantially. Thus, the salinity variations of intertidal habitats are very much greater than those in the open sea, and the potential variability at any height on the shore will be proportional to the length of emersion.

Emersion will also result in an immediate change in the CO_2-supply for intertidal plants. The absolute concentration of CO_2 in air is similar to that of molecular CO_2 in sea water (see Table 2.3), but the large buffering reservoir of bicarbonate and carbonate ions is no longer present, and the gas must be dissolved in a surface film of water on the thallus before it is available for photosynthesis. The exact effect of this change in CO_2-supply on the photosynthetic rates of intertidal algae has yet to be fully evaluated (see p. 126), but it seems probable that CO_2 will become limiting for photosynthesis more often when the plants are emersed than when they are submerged. CO_2-limitation must certainly occur in small rock pools during emersion, since high temperatures combine with photosynthesis to reduce the concentration of dissolved CO_2 to very low levels, and pH-values as high as 9.9 have been reported.[233] The supply of other plant nutrients (especially nitrogen and phosphorus) is also more restricted in intertidal habitats than in the open sea simply because marine algae have no roots and can take up mineral salts only when they are covered by water. The rate of nutrient uptake seems to be important for plants growing high on the shore. Photosynthesis is possible during prolonged emersion in cool, humid weather, but growth may be limited by the time available for absorbing the necessary mineral salts (see p. 128).

When plants are emersed on the shore, they are exposed to full daylight, but the change in irradiance on emersion is not as sudden as the changes that occur in other environmental factors. The irradiance gradually increases as the water level falls, and then gradually decreases as the tide comes in again. The combined effects on irradiance of the changing depth of water above the plant and the different periods of emersion at different heights on the shore can be calculated and expressed as a percentage of the total light at the surface, integrated over a complete tidal cycle of 12.5 hours (Fig. 2.13). The irradiance decreases almost linearly with depth, whereas, for continually submerged plants, the decrease with depth is exponential and much more rapid (see Fig. 2.5). At any given height on the shore, the irradiance varies through the spring/neap cycle of the tides. The upper shore receives more light during neap tides than during spring tides, but the lower shore (from about 1 m below MTL) receives more light during spring tides (Fig. 2.13). These calculations take no account of the diurnal variations in surface irradiance, and so the timing of the low waters in either the spring or the neap series has an important effect on the absolute irradiance received. If low water springs occur at around midday, so that neaps occur at 6 a.m. and 6 p.m., all parts of the shore receive more light during spring than during neap tides.

Wave action and exposure

Tides are not the only form of water movement that occurs in the sea, and waves often exert an influence on shores that is as great as, or even greater than, that of the tides. As far as intertidal plants are concerned, some of the effects of wave action are beneficial, and some are harmful. The splash from breaking waves keeps plants moist and cool for some time after the tide has receded, and effectively raises the water level above the predicted tidal height and reduces the emersion time at any given height on the shore. This applies even to levels that are not normally reached by the tide, so that some marine species are able to live above the highest level that is ever completely submerged by sea water. The height reached by the splash from a particular wave depends not only on the size and the energy of the wave, which are determined largely by the wind speed, the fetch and the depth of water adjacent to the shore, but also on the slope of the shore that the wave breaks against. This height is much more variable, therefore, along an irregular rocky shoreline than along a sandy beach, or along a concrete wall or break-water. However, the average height reached by the splash from the waves (the 'swash') above the predicted tidal level can be estimated, and this has been used as a measure of the effect of wave action on certain aspects of the intertidal environment.

Waves may also exert a beneficial effect on plant growth by increasing the speed of water movement. This reduces the thickness of the still boundary layer that surrounds all solid surfaces in a fluid medium, and thus reduces the resistance to diffusion that this boundary imposes. The sea is, of course, rarely completely still, and the increased water movement caused by the waves will not necessarily result in a significant decrease in the thickness of the boundary layer, but the growth of most aquatic macrophytes increases with current velocity,[41] and estimates of this velocity in the intertidal may well be useful. Since water movement is so variable on a wave-battered shore, some integrated measure over a period of a day or more is probably more valuable than readings at isolated instants in time. The simplest — and possibly most effective — answer to this problem is to measure the rate of dissolution of plaster of Paris balls[173] or 'clod-cards'[55] placed in the marine environment.

The most obvious effect of wave action on intertidal habitats, however, is the sheer physical damage that is caused both to the organisms and to their substrate. This damage can be attributed to three main effects[112]

1. Abrasion — caused either by particles suspended in the water, or by flexible plants being lashed against rock surfaces.

2. Pressure: pockets of air trapped in a breaker may be strongly compressed (up to 355 kg mm^{-2} has been recorded) but, since marine plants contain few air spaces and are largely incompressible, little damage is probably caused by such pressures.

3. Drag: it is the physical removal of an organism from its anchorage that constitutes the most drastic and irreparable damage, and measurements of the maximum drag forces operating in an intertidal habitat are of great ecological interest.

Jones and Demetropoulos[112] investigated these forces using a simple 'dynamometer', consisting of a disc (4 cm in diameter) attached to a spring balance. The balances were anchored to the rocks at about MTL at a series of stations along the north shore of a bay, and the maximum forces were recorded at intervals through the winter (October – April). The results (Fig. 2.14) show that the drag forces near the head of the bay are about one-fifth of those on the headland, which is fully exposed to prevailing south-westerlies with a fetch of 6000 miles (9700 km). A comparison of wave heights (calculated from wind speed and direction) with dynamometer readings at the different stations (Fig. 2.15) shows that, for all stations, the drag forces are proportional to wave height for small waves, but become independent of wave height as the size of the waves increases. In sites nearest to the head of the bay (low numbers, see Fig. 2.14), drag force becomes independent of wave height at smaller wave sizes than in the more exposed sites near the headland. This is because the water is shallower towards the head of the bay, and the waves break sooner and dissipate some of their energy before reaching the shore. Although all of these measurements were made at MTL, meters fixed at a series of vertical heights above MTL showed that the maximum drag forces were much the same all the way up to the mean level of high spring tides. The maximum wave impact would be expected to decrease, however, towards the lower regions of the intertidal, since the crests of the larger waves always break above the level of the lowest tides.

The results obtained with these dynamometers correlate well with subjective judgements of the relative exposure to — or shelter from — wave action at each of the sites, and some marine ecologists believe that subjective estimates of exposure are just as useful as exact measurements of particular physical properties. The major problem with subjective estimates of exposure, however, is that it is impossible to make a detailed comparison between estimates made by different observers in different parts of the world. An alternative approach has been to devise an index of exposure based on the aspect of the shore and the direction and strength of the winds. For example, Moore[162] suggested a wave exposure scale based on the percentage of days in which any wind blows directly onto the shore, and Baardseth[8] counted the number of 9° compass sectors with an unimpeded fetch of at

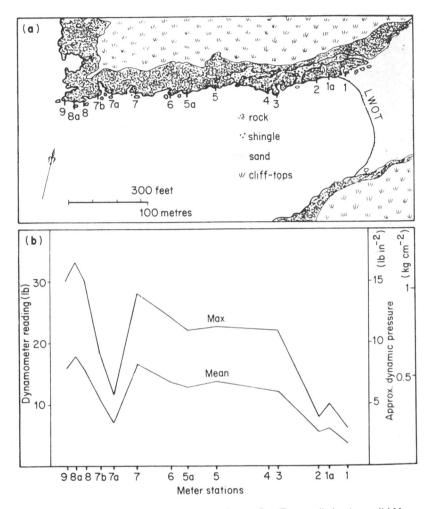

Fig. 2.14 **(a)** Positions of dynamometer stations at Port Trecastell, Anglesey. **(b)** Mean and maximum readings obtained at each meter station, plotted according to their positions on the map.[112]

least 7.5 km from the shore. The latter index gave a reasonably good correlation with the dynamometer readings in Fig. 2.14,[216] but there have, unfortunately, been few other comparative studies of different physical measurements of exposure.

Most shore ecologists, indeed, seem to take the view that no physical measurement of exposure can estimate more than one aspect of this factor, whereas intertidal organisms respond to several different aspects simultaneously. Therefore, the best integrated measure of the degree of exposure of a shore may perhaps be obtained by observing the responses of the

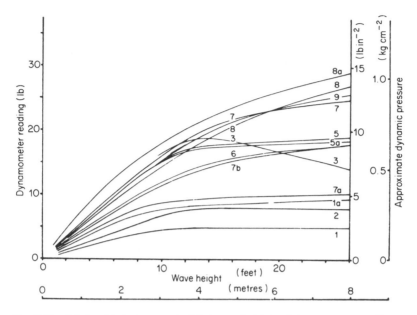

Fig. 2.15 Relationship between calculated wave height and dynamometer readings at different stations, numbered as in Fig. 2.14.[112]

organisms themselves, and a number of biological exposure scales have been devised on this basis.[50, 135] These have two major disadvantages. They can be used only within the geographical range of the species included in the assessment, and they can all too easily lead into deceptive, circular arguments (e.g. 'species A occurs only on exposed shores' — but these shores have been described as 'exposed' because of the presence of species A). The problem of how to measure this important aspect of the intertidal environment still awaits a satisfactory solution.

3

Photosynthesis in the Sea

In the more ecology-conscious world that has developed since the late 1960s, it has become almost a cliché to observe that 'all flesh is grass' or — to put it more scientifically — that the whole of the biosphere depends on the conversion of radiant solar energy to chemical energy that occurs in the process of photosynthesis. What is sometimes overlooked is that, although photosynthesis provides chemical energy and carbon skeletons for biosynthesis (both in the form of carbohydrates), it does nothing to provide other essential requirements, such as nitrogen and sulphur for cell proteins, phosphorus for nucleic acids and membrane lipids, and the many other elements that are necessary for the final structure of enzymes, pigments and cell walls. Since the supply of these elements may not always match the supply of energy and organic carbon provided by photosynthesis, there is not always a direct relationship between photosynthesis and plant growth. This is particularly true of many marine plants growing in their natural habitat, and the separate discussion of photosynthesis in the present chapter and of growth in Chapter 4 emphasizes the differential effects of environmental factors on the two processes. The primary productivity of an ecosystem, whether it is defined in terms of energy fixation (i.e. kJ m^{-2} yr^{-1}) or change in biomass (i.e. kg C m^{-2} yr^{-1}), depends on the rate of photosynthesis per unit area of surface, and this is equal to the photosynthesis per unit of biomass multiplied by the amount of biomass per unit area (or the 'standing crop'). Since the latter is a result of the growth processes considered in Chapter 4, the discussion of primary production is also included in the next chapter.

THE PHYSIOLOGY OF PHOTOSYNTHESIS: A REMINDER

Although certain species of marine algae have sometimes been in the front line of research into the basic physiology of photosynthesis, they have usually achieved this distinction because of their physiological peculiarities or their experimental convenience, and not because they were specifically marine. Such work is not, therefore, discussed in detail here; the main emphasis is placed on those aspects of the physiology of photosynthesis which are characteristic of marine plants, and which contribute to an understanding of their ecology. This section provides a brief description of the terms and concepts that appear in the subsequent discussion, but it is no substitute for more detailed accounts of the physiology and biochemistry of photosynthesis (see for example[38]).

Early in this century, F. F. Blackman deduced that, since changes in temperature and CO_2 concentration affected the light-saturated rate of photosynthesis (or P_{max}) but had no effect on the initial slope of typical photosynthesis versus irradiance ('P vs I') curves (see Fig. 3.1), photosynthesis must consist of two distinct processes: a photochemical process (or 'light reaction') influenced primarily by irradiance and chlorophyll concentration; and an enzymic, chemical process (the 'dark reaction') which, like all chemical reactions, is affected by substrate (i.e. CO_2) concentration and by temperature. These two processes are linked by metabolic intermediates which cycle between them. When the overall rate of photosynthesis is light-limited, the 'dark reaction' is working below its maximum rate because these intermediates are not being supplied fast enough by the light reaction. At light saturation, however, the dark reaction is working at full capacity, and it is the rate of return of the intermediates that prevents the rate of the light reaction from increasing as the irradiance increases further. These intermediates are now known to be the nucleotides NADP and ATP (see Fig. 3.2).

Of the various pigments which have been found to be involved in photosynthesis, only chlorophyll a is common to all photosynthetic organisms

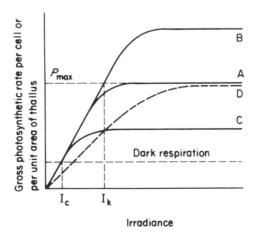

Fig. 3.1 Photosynthesis versus irradiance ('P vs I') curves under different conditions. If curve **A** represents measurements made under standard conditions, curve **B** shows the effects of increased temperature, curve **C** the effects of decreased CO_2 concentration, and curve **D** the effects of decreased chlorophyll per cell or per unit area. I_c: light compensation point for curves **A**, **B** and **C**; I_k: light saturation point for curve **A**.

which liberate oxygen. Chlorophyll *a* was soon recognized, therefore, to be
the main pigment in photosynthesis, and all other photosynthetic pigments
in plants are considered to be ***accessory pigments***. The fluorescence
spectra from whole plants or from isolated chloroplasts show that most of
the energy absorbed by the accessory pigments is transferred to chlorophyll
a. This transfer is possible because chlorophyll *a* absorbs at a longer wavelength than any of the other pigments, and so requires less energy for excitation. Even chlorophyll *a* itself, inside the chloroplast, exists in a number of
different forms, all of which have their maximal absorption at slightly
different wavelengths between 662 and 700 nm, and absorbed energy can be
transferred in sequence from the short-wavelength to the longer-wavelength
forms. The pigments in the plant appear to be organized into two distinct
systems. The bulk of the accessory pigments and the shorter-wavelength
forms of chlorophyll *a* are found in Photosystem II (PS II), whereas
Photosystem I (PS I) consists entirely of chlorophyll *a*, with a higher
proportion of long-wavelength forms than in PS II. Each reaction centre of
PS I contains a single molecule of a form of chlorophyll *a* that absorbs at 700
nm, called P700, and this acts as a trap for all of the energy absorbed by this
photosystem. PS II is thought to have a similar energy trap, P680 – 690, but
this has not yet been isolated. The two photosystems provide the energy for
two distinct photochemical reactions, which are linked in series to produce
ATP and $NADPH_2$, as shown in Fig. 3.2.

The distribution of the pigments between the two photosystems and
the existence of two distinct light reactions provide an explanation of
the ***enhancement effect***. When plants are irradiated with light that is
absorbed only by PS I (i.e. wavelengths longer than about 680 nm in green

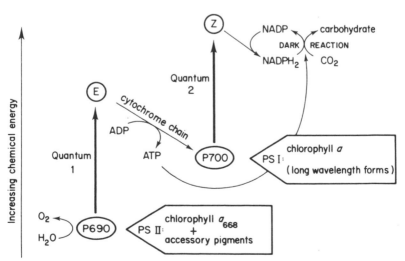

Fig. 3.2 Generalized scheme for electron transport in the light reactions of
photosynthesis.

algae and higher plants), the photosynthesis yield is very poor. If the plants are simultaneously irradiated with shorter-wavelength light, that can be absorbed by PS II, however, the yield is greater than the sum of the yields of the two wavelengths used separately (i.e. the yield is 'enhanced'). Light absorbed by PS I can be used in only the second of the light reactions, which reduces NADP and oxidizes the cytochrome chain, and cannot be transferred to PS II, which reduces the cytochromes with electrons removed from water (Fig. 3.2). Some of the energy absorbed by PS II, however, can be transferred to PS I, so that both light reactions can operate, but maximum efficiency is achieved only when both photosystems are irradiated together.

PIGMENTS AND PHOTOSYNTHESIS IN MARINE PLANTS

A superficial examination of the seaweeds of a rocky shore is enough to establish an important distinction between marine plants and terrestrial plants. The vegetative parts of the vast majority of the land flora are green, but plants growing in the sea are far more varied in colour. The biological importance of the differences in colour between different algal species was recognized in the early classification of these plants (see p. 2), and more recent physiological work has established that the characteristic colours of different algal groups represent different combinations of photosynthetic pigments (see Table 1.2). Since all higher plants, together with green algae, contain chlorophyll *b* as the main accessory pigment, the data in Table 1.1 indicate that this pigment occurs in over 98% of photosynthetic species on land and in fresh water, whereas less than 13% of marine species contain it. Since the Chrysophyta, Rhodophyta, Phaeophyta and Pyrrhophyta are all at least as well represented as the Chlorophyta in the marine flora (Table 1.1), it is clear that all of the accessory pigments listed in Table 1.2 make a significant contribution to marine photosynthesis. In addition, the light climate in the sea is far more varied — both in quality and in quantity — than that on land (see pp. 14–18), and it is reasonable to suggest that there may be a connection between the variability of pigment composition and the variability of light climate in the sea. This idea is examined towards the end of this section.

Pigment composition

The main accessory pigments of each of the major groups of marine algae have already been described (p. 2, Table 1.2), but no indication of the quantitative importance of these pigments relative to chlorophyll *a* has been given. Marine green algae seem to contain more chlorophyll *b* per unit of chlorophyll *a* than do either higher plants or freshwater green algae. The accessory pigment ranges from 44 to 72% of the chlorophyll *a* content in species as diverse as *Dunaliella, Ulva* and *Codium* (see Table 1.3 for the morphology of these plants), whereas values reported for higher plants range from 20 to 35%.[174] Chlorophyll *c* exhibits a similar range of concentration, amounting to at least 20% of the chlorophyll *a* content in most brown algae, and reaching 50-60% in some diatoms and dinoflagellates.[105] Two chemically-distinct forms of chlorophyll c occur (c_1 and c_2), which have slightly different distributions among the various groups of algae (see Table 1.2), but the absorption properties of the two forms are so similar that little physiological significance can be attached at present to the distinction between them. In fact, the physiological contribution of chlorophyll c is usually overshadowed by a second accessory pigment — either fucoxanthin or peridinin. The fucoxanthin: chlorophyll a ratio is 0.3-0.4 in brown algae[205] and 0.5-0.6 in diatoms and, in dinoflagellates, peridinin occurs in much the same abundance relative to chlorophyll a.[248]

All of the accessory pigments discussed so far resemble chlorophyll a in being insoluble in water and extractable only with organic solvents, such as acetone or methanol. Since similar extraction methods are used, it is possible to obtain a meaningful comparison of accessory pigment and chlorophyll a concentrations, even if the extraction is not complete, or extraction losses occur. The main accessory pigments of the red and blue-green algae, how-ever, are phycobilins, and these are not soluble in organic solvents. They can often be extracted from plant tissues quite easily with water, but the extrac-tion is rarely complete, and the pigments are difficult to separate from the protein molecules to which they are bound. For these reasons, it is not easy to estimate either the absolute concentration of phycobilin pigments in algal tissues, or the ratio of phycobilin to chlorophyll a concentration. A less exact, but physiologically meaningful, estimate of the relative importance of phycobilins and chlorophyll a in these plants is often obtained by measuring the absorption spectrum of intact plants, and comparing the peak heights attributable to phycoerythrin or phycocyanin with that of chlorophyll a (see Fig. 3.4e, f).

In common with higher plants, all algae contain small concentrations of other carotenoid pigments, such as β-carotene and a selection from the wide variety of different xanthophylls. These do not appear to transfer much of their absorbed energy to chlorophyll a, and their primary function is thought to be the protection of chlorophyll from photo-oxidation by very high irradiances (see p. 57). The chemical structure and *in vitro* absorption properties of all of these algal pigments are given by Stewart.[239]

Pigment-protein complexes

A clearer picture of the way in which light energy absorbed by chlorophyll a and the accessory pigments is transferred to the reaction centres of the two photosystems is beginning to emerge from the recent isolation and charac-terization of pigment-protein complexes from chloroplasts. The insolubility of chlorophylls and carotenoids in water has meant that these pigments have traditionally been extracted with organic solvents but, since these dis-solve the chloroplast membranes on which the pigments are located, they destroy the *in vivo* organization of the pigment molecules. It has long been recognized that this resulted in substantial changes in the absorption properties of many pigments. For example, fucoxanthin absorbs very little light at wavelengths longer than 500 nm when it is extracted in organic solvents but, in diatom and brown algal cells, the pigment absorbs mainly between 500 and 590 nm. Most of the chlorophylls show similar, if less striking, shifts of their absorption maxima to shorter wavelengths on extraction, and the various forms of chlorophyll a that are observed *in vivo* (see p. 45) can no longer be distinguished. It is for this reason that the *in vitro* absorption properties of the different algal pigments are not given here. They have very little physiological significance.

If chloroplast membranes are broken up gently by detergent action, rather than with organic solvents, the pigments remain attached to the protein components of the membranes, and a number of distinct pigment-protein complexes can be separated. All photosynthetic plants appear to contain a complex consisting of P700, chlorophyll a and protein, with about 40 chlorophyll a molecules to every molecule of P700.[3] This complex carries out photochemical reactions *in vitro* and is thought to represent the reaction centre of PS I. It contains $10-20\%$ of the total chlorophyll of the plant, and a further $40-60\%$ of the total chlorophyll is located, together with almost all of the accessory chlorophylls and carotenoids, in chlorophyll a-accessory pigment-protein complexes, whose composition varies from one algal group to another (Table 3.1). The chlorophyll a/b-protein complex of higher plants was initially thought to be PS II, but it shows no photochemical activity, and all of the complexes listed in Table 3.1 are now interpreted as light-harvesting components of the photosynthetic apparatus, which may contribute absorbed energy to either of the photosystems. The reaction centre of PS II has not yet been isolated. The absorption properties of these light-harvesting complexes resemble those of the intact plant from which they were obtained, and the emission and excitation of fluorescence (Fig. 3.3) show that energy transfer between accessory pigments and chlorophyll a occurs in the complexes with the same efficiency as in the intact plant. The light-harvesting complexes are the most variable component of the photosynthetic apparatus, and show a greater response to changes in environmental factors than does the P700-chlorophyll a-protein.

Table 3.1 Light-harvesting pigment-protein complexes of marine algae.

| Algal group | Complex | Ratio of accessory:chl a | | |
		In complex	In thallus	Reference
Green algae	Chl a/b-protein	$0.8-1.1$	$0.4-0.7$	174
Brown algae	Fucoxanthin-chl a/c_2-protein Chl $a/c_1 + c_2$-protein	c_2: 0.5 c : 0.33	$0.2-0.3$	3
Diatoms	Fucoxanthin-chl a/c-protein	f : 0.5 c : 0.5		97
Dinoflagellates	Peridinin-chl a-protein	4.0	$0.7-1.1$	194
Cryptomonads Red algae Blue-green algae	Phycobiliproteins	No chl a	$0.7-2.5$	81

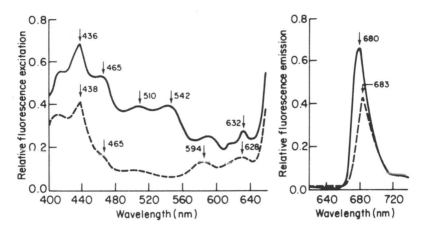

Fig. 3.3 Action spectrum for the excitation of fluorescence, and the spectrum of emitted fluorescence, for the light-harvesting protein complexes isolated from the brown seaweed *Acrocarpia paniculata* (Fucales).[3] Continuous line: fucoxanthin-chlorophyll a/c_2-protein; broken line: chlorophyll $a/c_1 + c_2$-protein.

The light-harvesting complexes of red and blue-green algae consist entirely of phycobilin-protein complexes, which are located outside the thylakoid membranes in spherical granules known as **phycobilisomes.** The various phycobilin pigments are thought to be arranged in layers in these granules, with phycoerythrin on the outside, phycocyanin in the middle layer, and allophycocyanin at the centre, where it is in contact with the thylakoid membranes containing chlorophyll a.[81] This arrangement facilitates energy transfer, since the pigments occur in the same sequence as the wavelengths of the radiation which they absorb. P700-chlorophyll a-protein complexes, similar to those of other plants, have been isolated from these algae, but the apparent absence of chlorophyll a from the light-harvesting complexes may have important physiological implications (see below).

Action spectra for photosynthesis and their interpretation

The overall contribution of each pigment to the photosynthesis of intact plants can best be judged by comparing the *in vivo* absorption spectrum with the action spectrum for photosynthesis. An action spectrum represents the amount or rate of a photobiological process that occurs in monochromatic light of equal photon irradiance at different wavelengths. If the energy absorbed by all the pigments in a plant were utilized with equal efficiency in photosynthesis, the absorption spectrum and the action spectrum for photosynthesis should be very similar. Discrepancies between absorption spectra and action spectra suggest that the energy absorbed by certain pigments is not utilized with the same efficiency as that absorbed by other pigments.

These two types of spectrum are shown for six species of marine algae in Fig. 3.4. Two of the species are unicellular — a diatom (Fig. 3.4c) and a dinoflagellate (Fig. 3.4b) — and the rest have flat parenchymatous thalli of various thicknesses (see Table 1.3). In the green and brown-coloured algae (Fig. 3.4a–d), there is a close correlation between absorption spectra and action spectra, but all three species containing chlorophyll c (Fig. 3.4b–d) show greater absorption and photosynthesis in green light (500–600 nm) than the green alga (Fig. 3.4a). The absorption spectrum for *Laminaria* (Fig. 3.4d) is much flatter than those for the other species. This is because a mature kelp blade was examined, and this was thick enough to absorb almost

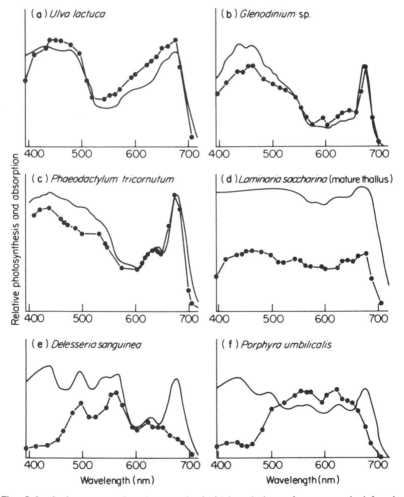

Fig. 3.4 Action spectra for photosynthesis (●) and absorption spectra (—) for six species of marine algae. ((a), (d)–(f) Unpublished data of K. Lüning; (b) Prézelin *et al.*;[195] (c) Mann, J.E. and Myers, J. (1968). *Plant Physiology, Lancaster*, **43**, 1991–5.)

all of the visible light, regardless of wavelength. The flat action spectrum for this species indicates that almost all wavelengths produce the same amount of photosynthesis, and that the 'gap' in the absorption spectrum between 500 and 650 nm (see Fig. 3.4a – c) can be 'filled' by increasing the concentration of the photosynthetic pigments, as well as by altering their composition. The equally thick thallus of the green alga *Codium* also has a flat absorption spectrum (i.e. it is optically black), and the action spectrum is almost the same as in *Laminaria*, in spite of the different pigments present.

The absorption spectra of the two red algae in Fig. 3.4 differ from one another because the balance between the two phycobilin pigments, phycoerythrin and phycocyanin, is quite different in the two species. In *Delesseria* (Fig. 3.4e), which is a bright red seaweed, phycoerythrin is the dominant pigment and there is little phycocyanin but, in *Porphyra* (Fig. 3.4f), the phycocyanin concentration is slightly greater than that of phycoerythrin, and the thallus is reddish-brown. In both species, however, there are large discrepancies between the absorption spectrum and the action spectrum, and it appears that light absorbed by chlorophyll *a* alone (at both ends of the spectrum) produces very little photosynthesis. If these plants are irradiated with red or blue light *together with* an intermediate wavelength that is absorbed by one of the phycobilin pigments, enhancement of photosynthesis (see p. 45) occurs, and the efficiency of utilization of the red or blue light is substantially increased. The poor photosynthesis of red (and blue-green) algae in light which is absorbed by chlorophyll *a* alone cannot be completely explained by saying that both photosystems must absorb light simultaneously (see p. 46), since PS II is generally thought to contain short-wavelength forms of chlorophyll *a*, and green and brown algae achieve efficient photosynthesis in wavelengths (670 – 680 nm) which are not absorbed by any of their accessory pigments (Fig. 3.4a – d). The full explanation is probably related to the apparent absence of chlorophyll *a* from the light-harvesting pigment-protein complexes of phycobilin-containing algae (see p. 49).

Action spectra thus provide valuable information about the physiology of photosynthesis in different plants, but they can also be useful in understanding the distribution and performance of plants in the sea, where the spectral distribution of underwater light is so variable (see p. 14). The total photosynthesis of a plant in any light field can be calculated from the spectral distribution of the light and the action spectrum for photosynthesis. The photon irradiance at each wavelength is simply multiplied by the photosynthesis per photon at the same wavelength, and the sum of these products for all wavelengths between 400 and 700 nm provides an estimate of the total photosynthesis. This calculation assumes that the effects of each wavelength on photosynthesis are independent of all other wavelengths, but the enhancement effect (see p. 45) indicates that this is not always true — at least, under laboratory conditions using monochromatic light. The contribution of enhancement to the total photosynthesis of various marine algae in 'white' light, similar in quality to underwater light, has been assessed by comparing the measured rate of photosynthesis with that calculated on the assumption that no enhancement occurs.[61] Measured photosynthesis was about 10–20% higher than the calculated values for red algae in spectra typical of coastal waters, but very little enhancement could be detected in green or brown algae. These results suggest that enhancement has only a limited influence on photosynthesis in the sea, probably because the wavelengths which can be most strongly enhanced in laboratory conditions (i.e. 400–500 and > 650 nm for red algae; > 680 nm for green and brown algae) are all poorly represented in underwater light (see Fig. 2.3). This also means that the calculations outlined above provide a valid approach to the estimation of photosynthetic rates in the sea.

Variation of pigment composition with water depth

This topic has been strongly influenced for nearly a century by the idea — first proposed by Engelmann in 1883[69] — that different groups of marine algae dominate the benthic vegetation at different depths because their pigment composition adapts them for absorption, and hence photosynthesis, in the light quality that prevails at that depth. This concept, popularly known as 'chromatic adaptation' (although it would be better to reserve this term for phenotypic changes in pigment composition; see p. 54), has been used to explain why blue-green and green algae were dominant only in the upper part of the intertidal, brown algae were dominant through the middle and lower intertidal and the upper part of the subtidal, and red algae were dominant towards the lower limit of the photic zone. As the light became progressively greener with depth (Fig. 2.3), it was argued, so the possession of, first, fucoxanthin and then phycoerythrin became more advantageous. It has often been pointed out that this theory is based on extremely superficial generalizations about the vertical distribution of benthic marine algae, and that representatives of most of the major groups can be found at most depths, but the theory continues to be widely quoted, and it often appears in elementary texts as a simple and elegant example of the physiological adaptation of plants to their environment. It is an uncomfortable fact of scientific life, however, that the simplicity and elegance of a theory is no guarantee of its validity, and this one needs to be subjected to a more careful scrutiny than it has usually received in the past.

In order to demonstrate that light quality is a significant factor in the vertical distribution of benthic algae, it is necessary to show that, for example, the photosynthesis of green algae is greater than that of red algae in light qualities typical of surface conditions (or vice versa in deep-water light qualities) *and* that the observed differences in photosynthetic performance are due to the pigment composition of the algae, and not to some other attribute of the photosynthetic apparatus. For example, the photosynthesis of mature fronds of *Laminaria* has been shown to be virtually independent of wavelength (Fig. 3.4d) and, therefore, the changes in light quality that occur with depth can have little effect on the photosynthesis of this plant. The ecological dominance of the kelps (*Laminaria* and its relatives) in the upper subtidal zone of most of the world's temperate coasts (see Chapter 7) cannot, therefore, be attributed to the specific adaptation of their pigment composition to the light quality in that depth zone.

As the *Laminaria* sporophyte develops and the blade thickens, the pigment concentration per unit area of thallus increases and fills in the 'hole' in the absorption spectrum, so that both the absorption spectrum and the action spectrum get flatter (Fig. 3.4d). A similar result is obtained with *any* action spectrum as the incident irradiance is increased. An action spectrum must be measured at relatively low irradiances, because wavelengths which produce faster rates of photosynthesis also saturate photosynthesis at lower irradiances. Increasing the irradiance beyond this level does not increase photosynthesis at the peaks of the action spectrum, but causes an increase at other wavelengths. Thus, the photosynthetic rates at all wavelengths gradually 'catch up' with those at the peaks and, in completely saturating irradiances, photosynthesis is independent of both light quality and pigment composition. All marine plants that grow in the intertidal zone or just below ELWS are frequently exposed to irradiances above saturation (see Table 6.4 for typical saturating irradiances) and, therefore, Engelmann's hypothesis cannot be applied to the vertical distribution of algae near the sea surface.

The possibility remains that light quality contributes to the control of algal distribution in deep water. At the lower limit of the photic zone, light is (by definition) the limiting factor, and many of the algae have thin thalli. Therefore, photosynthesis can be expected to be influenced by light quality in deep-water habitats. Is there any evidence, then, that the deepest-growing algae in each water type are those with the best pigment composition for photosynthesis in the spectral distribution at that depth? A comparison of the theoretical photosynthesis (calculated as described on p. 51) of four algal species at the lower limit of the photic zone in each water type (Table 3.2) shows that *Delesseria* (see p. 51, Fig. 3.4e) has the highest photosynthesis in the middle four water types (oceanic III to coastal 5). This result agrees with ecological observations, since *Delesseria* is normally found at greater depths than the other species in the table. In clearer oceanic water (types I and II), however, the green and brown algae show substantially higher photosynthesis than the two red algae, and this suggests that the former groups should be able to grow at greater depths than red algae in such waters. This is not observed: brown and green algae are found in relatively

Table 3.2 Predicted photosynthesis per incident photon, calculated from the action spectrum for photosynthesis of four species of algae and the spectral distribution of the light at the lower limit of the photic zone in all water types.[60]

| Water type | I | II | III | 1 | 3 | 5 | 7 | 9 |
Depth (m)	175	90	55	50	33	20	14	10.5
Ulva lactuca	1.3	1.2	1.0	0.7	0.7	0.8	0.8	0.8
Laminaria saccharina	1.2	1.2	1.2	1.1	1.1	1.1	1.0	1.0
Porphyra umbilicalis	0.5	0.8	1.1	1.5	1.5	1.5	1.5	1.6
Delesseria sanguinea	0.9	1.1	1.4	1.6	1.6	1.6	1.4	1.3

deep oceanic waters, but the deepest algae still tend to be red algae. Similarly, in turbid coastal waters (types 7 and 9, Table 3.2), the predicted photosynthesis suggests that *Porphyra* should replace *Delesseria* towards the bottom of the photic zone but, again, this is never observed in the sea. *Porphyra* and other red algae with a high phycocyanin content (Fig. 3.4f) are typically intertidal forms, and are not found at depth. Thus predictions based on Engelmann's hypothesis do not always coincide with actual distributions, and there seems to be little convincing evidence that light quality influences the distribution of algal groups through their photosynthetic pigment composition.

The differences in pigment composition discussed so far have all been differences between species, and can be described as genetically fixed, or **genotypic,** variations. However, the pigment composition of individual plants has been shown to change within a few days when they are transferred to different depths. These **phenotypic** variations in pigment composition could arise in response to the changes in light quality that occur with depth, but they could also be caused by the decrease in irradiance. After 7 days at 10 m, the red algae *Porphyra* and *Chondrus* had higher phycoerythrin: chlorophyll *a* ratios than plants at 1 m (Table 3.3), and this phenotypic change in pigment composition could be regarded as a chromatic adaptation, since the light at 10 m was greener than that at 1 m. A similar experiment with brown algae, however, showed that the fucoxanthin: chlorophyll *a* ratio *decreased* at depth (*Ascophyllum, Fucus,* Table 3.3), which is the reverse of what would be expected if chromatic adaptation were occurring. A comparison of the pigment composition of 'sun' and 'shade' plants with those at different depths showed that, in all algal groups, the changes that occurred in response to a decrease in irradiance (i.e. in shade plants) were qualitatively similar to changes in response to an increase in depth (Table 3.3). These results clearly suggest that phenotypic variations in pigment composition with depth are controlled by the irradiance and not by the

Table 3.3 Accessory pigment; chlorophyll *a* ratios in marine algae grown at different depths and in different irradiances at the same depth.[204, 205]

Species	Water depth* Shallow	Deep	Intertidal plants 'Sun'	'Shade'	Accessory pigment
Porphyra umbilicalis	0.47	0.76	0.60	0.73	Phycoerythrin
Chondrus crispus	0.59	0.89	0.50†	0.83†	
Ascophyllum nodosum	0.42	0.34	0.42	0.27	Fucoxanthin
Fucus vesiculosus	0.42	0.34	0.43	0.32	
Ulva lactuca	0.44	0.67	0.53	0.62	Chlorophyll *b*
Codium fragile	0.63	0.67	0.68	0.72	

* Plants were suspended for 7 days at 1 and 10 m (red and green algae) or at 0 and 4 m (brown algae)
† Plants at 3 m; 'sun' plants exposed to high irradiances by removing canopy species[210]

quality of the light; laboratory studies of marine algae grown at different irradiances support this conclusion (see below).

These and other arguments for discarding the theory of 'chromatic adaptation' are reviewed by Ramus,[203] and it looks as though the centenary of Engelmann's hypothesis may not, after all, be celebrated. However, there has always been an alternative to this theory. Oltmanns, another remarkable pioneer in the physiological ecology of algae, pointed out in 1892 that the change in light quality that occurs with increasing depth in the sea is invariably accompanied by a decrease in irradiance,[184] and that, therefore, it is impossible to distinguish between the effects of these two factors simply by studying the vertical distribution of benthic plants. In submarine caves, however, it is possible to find a gradient of irradiance without concurrent changes in light quality, and here the sequence of the species is much the same as that observed with increasing depth. Oltmanns concluded, therefore, that irradiance was more important than spectral distribution in determining the depths at which different species are found. This hypothesis is best tested by examining the long-term effects of growth at different irradiances on both the pigment composition and the photosynthesis of marine plants.

Effects of growth irradiance on the composition and concentration of pigments

Higher plants typical of shaded habitats ('shade plants') show differences in leaf structure, pigment composition and photosynthetic characteristics (especially compensation and saturation points) when compared with plants of open habitats ('sun plants').[18] As far as pigments are concerned, shade plants tend to contain more chlorophyll per unit weight of leaves, and have a higher chlorophyll $b:a$ ratio than sun plants; these differences may occur within a single species, when plants are grown in different irradiances (i.e. they may be phenotypic differences). Recent studies of seaweeds from shaded and sunlit positions in the intertidal zone have shown that similar differences in chlorophyll concentration can occur within species of all three major groups, and that the chlorophyll $b:a$ ratio in green seaweeds responds to irradiance in the same way as in higher plants (Table 3.3). In addition to these surveys in the field, laboratory studies of pigment composition in algae cultured at different irradiances have shown that a decrease in irradiance tends to increase the chlorophyll content per cell, and to change the accessory pigment:chlorophyll a ratio. Table 3.4 summarizes the effects of irradiance on the pigment composition of six marine algae representing all of the major groups of unicellular and multicellular forms. Since these results were obtained in six separate investigations at different irradiances and different temperatures, and some of the pigment concentrations were expressed per cell, some per unit area of thallus, and some per unit weight of wet or dry alga, it is impossible to make meaningful comparisons between the *absolute* pigment concentrations in different algae. The results are, therefore, presented as the percentage *change* in the concentration of each

Table 3.4 Effects of growth irradiance on the concentration of different photosynthetic pigments, and on the ratio of the major accessory pigment to chlorophyll *a*, in marine algae of different groups.

Alga	Group	Irradiance (μmol m^{-2} s^{-1}) High	Low	Percentage change from high to low irradiance Pigment concentrations Chl *a*	Chl *b*,*c*	Others	Accessory ratio
Ulva	green	sun	shade	+674	+774[1]	–	+17[1]
Sphacelaria	brown	240	8	+490	+257[2]	+275[3]	–36[3]
Phaeodactylum	diatom	300	30	+238	nm[2]	+63[3]	–21[3]
Amphidinium	dinoflag.	450	10	ns	+27[2]	+72[4]	+72[4]
Chroomonas	crypto.	135	5	+19	+6[2]	+54[5]	+32[5]
Griffithsia	red	200	1.6	nm	–	nm	+162[6]

Pigments: [1]chlorophyll *b*; [2]chlorophyll *c*; [3]fucoxanthin; [4]peridinin; [5]phycocyanin; [6]phycoerythrin; nm = not measured; ns = no significant change
References: *Ulva*,[204] *Sphacelaria*,[24] *Phaeodactylum*,[230] *Amphidinium*,[152] *Chroomonas*,[75] *Griffithsia*[250]

pigment, or in each pigment ratio, that occurred in response to the decrease in irradiance from 'high' to 'low' light, Relatively few algal species have been studied in this way, so that the evidence is somewhat fragmentary as yet, but the pattern seen in Table 3.4 is confirmed by most of the results available.

Except at very low irradiances, where light stress may cause a decrease in pigment concentration (e.g. the dinoflagellate *Gonyaulax*[196]), total pigment concentration tends to increase in all algae when the irradiance is decreased. All of the pigments in the plant contribute to this increase in most species, although chlorophyll *a* remains relatively constant in a few (e.g. *Amphidinium*, Table 3.4; some diatoms[116]). The different pigments rarely increase in concentration at the same rate, however, so that substantial changes occur in the ratios between them. Chlorophyll *b*, peridinin and the phycobilins all increase more rapidly than chlorophyll *a* in response to a decrease in irradiance, and the accessory pigment:chlorophyll *a* ratio increases (*Ulva*, *Amphidinium*, *Chroomonas*, *Griffithsia*; Table 3.4). This is similar to the response of higher plants, where the increased chlorophyll *b*:*a* ratio of shaded plants appears to be due to a specific increase in the light-harvesting component of the photosynthetic apparatus. Since this complex contains a higher chlorophyll *b*:*a* ratio than the whole plant (see Table 3.1), chlorophyll *b* increases more rapidly than chlorophyll *a*. This mechanism has not yet been demonstrated in a marine green alga, but comparable changes have been observed in a dinoflagellate and a red alga. In *Glenodinium*, it is the peridinin-chlorophyll *a*-protein complex which increases at low growth irradiances, while other components of the photosynthetic apparatus remain constant[193] and, in *Griffithsia*, reduced light causes a marked increase in the number of phycobilisomes per unit area of thylakoid membrane.[250] In view of these specific effects of low light levels on the light-harvesting complexes of higher plants and most algae, it is surprising that fucoxanthin does not behave in the same way as the other accessory pigments. In both brown algae and diatoms, the fucoxanthin:chlorophyll *a* ratio decreases as irradiance decreases (Tables 3.3, 3.4). This may mean that fucoxanthin performs two separate functions in the photosynthetic apparatus: a light-harvesting function at low irradiances, and a protective function at high irradiances. The action spectra for photosynthesis of *Phaeodactylum* cells grown in high and low light indicate that less of the energy absorbed by fucoxanthin in high-light-grown cells is transferred to chlorophyll *a* and used in photosynthesis than in low-light cells.[230] This suggests that, in high irradiances, some of the fucoxanthin is uncoupled from the photosynthetic apparatus, as would be expected if this pigment was protecting the chlorophyll from excessive irradiances, in much the same way that other carotenoids are thought to act in green plants (see p. 47).

The inhibitory effect of high irradiances on photosynthesis is particularly noticeable when the primary productivity of phytoplankton is measured *in situ* (see p. 85). The rates recorded at the surface are frequently lower than those a few metres below, and the magnitude of this surface inhibition in a large number of separate investigations has been shown[87] to be correlated with the extent to which the surface irradiance exceeded 200 μmol m^{-2} s^{-1}. This phenomenon of **photoinhibition** has proved very difficult to investigate in the laboratory because most artificial sources of intense light produce strong heating effects, and are deficient in the short wavelength radiation (near-UV and blue) which exerts the main inhibitory effects. Another problem is that the inhibition tends to increase with the length of exposure to high irradiances, so that measurements over 15- to 30-minute periods may underestimate the long-term effects of a particular light treatment.[87] One explanation of photoinhibition is that the excessive energy absorbed by the photosystems leads to the irreversible degradation of chlorophyll molecules by photo-oxidation. The dark reactions of photosynthesis may also be involved, however, since high irradiances of blue light have been shown to inhibit the activity of ribulose bisphosphate carboxylase[40] (see p. 65). Phytoplankton sampled from the surface of stratified water columns become photoinhibited more slowly than cells from deep water,[87] and this response may be related to an increase in the concentration of blue-absorbing carotenoid pigments similar to that observed in laboratory cultures (see above).

Chlorophyll concentration and photosynthesis

Growth irradiance is not the only factor that influences the chlorophyll content of marine plants. Chlorophyll concentration increases during development, both within the division cycle of an individual cell in the phytoplankton, and during the growth and maturation of a seaweed thallus. Nutrient availability may also affect the chlorophyll content, especially when elements such as nitrogen, iron or magnesium are in short supply. Any change in chlorophyll concentration, whatever its cause, is liable to change the P vs I curve of the plant, and some insight into the exact mechanism of the pigment changes can be obtained by comparing P vs I curves calculated in two different ways. The initial slope of a P vs I curve represents the light-limited phase of photosynthesis (Fig. 3.1), and its gradient will be affected by the efficiency with which the plant can absorb the limited light available. Increasing the chlorophyll concentration per cell, or per unit area of thallus, will clearly increase the percentage of the light that is absorbed, and so will increase this gradient, provided that the rate of photosynthesis is expressed per cell or per unit area (Fig. 3.5a,c). If photosynthesis is expressed per unit of chlorophyll *a*, however, the increased chlorophyll concentration will cancel out the increase in photosynthesis that it has achieved, and no change in the initial slope will be apparent (Fig. 3.5b, d). This prediction assumes that the light-limited rate of photosynthesis is proportional to the chlorophyll *a* concentration but, if the chlorophyll concentration is very high, some of the molecules may be shaded by other molecules, and the photosynthesis per unit of chlorophyll may decrease. This is most likely to occur in seaweeds with thick thalli (e.g. *Codium, Laminaria, Chondrus*). Another factor which may affect the apparent efficiency of the chlorophyll *a*

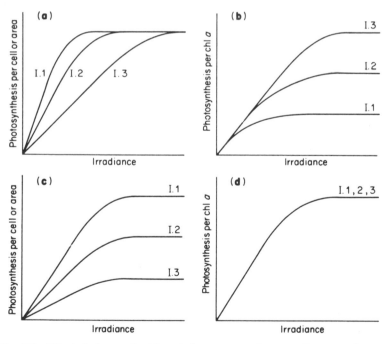

Fig. 3.5 Effect of changes in chlorophyll *a* concentration per cell or per unit area of thallus on P vs I curves expressed as both rate of photosynthesis per cell or per unit area, and rate of photosynthesis per unit of chlorophyll *a*. Chlorophyll concentration is assumed to be inversely related to growth irradiance (I), and I.1 < I.2 < I.3. **(a), (b)** Curves predicted if the *size* of existing PSUs changes in proportion to the chlorophyll concentration; **(c), (d)** curves predicted if the *number* of PSUs changes in proportion to the chlorophyll concentration.

is the proportion of accessory pigments present. If the accessory pigments increase more rapidly than the chlorophyll a, as happens when many algae are grown in low light (Table 3.4), the initial slope on a chlorophyll a basis is likely to increase slightly.

The chlorophyll in the thylakoid membranes is thought to be arranged in **photosynthetic units** (PSUs), each of which consists of 300 – 400 chlorophyll molecules associated with a single reaction centre.[38] Any increase in pigment concentration could, therefore, be achieved either by adding the extra pigment molecules to the existing PSUs, so as to increase the *size* of each unit without changing the number of reaction centres, or by building up complete new PSUs, with all the reaction centre components, so that the *number* of PSUs increases but their size remains constant. These two 'strategies' for increasing the pigments of a plant have similar implications for the initial slope of the P vs I curve, but will have quite different effects on the light-saturated rate of photosynthesis (P_{max}). Since P_{max} is related to the number of reaction centres available, an increase in the *size* of the PSUs will have no effect on the maximum rate of photosynthesis per unit area, and so P vs I curves for plants with different chlorophyll concentrations will have different initial slopes, but will all saturate at the same value (Fig. 3.5a). When photosynthesis is expressed on a chlorophyll basis, however, P_{max} will decrease as the chlorophyll concentration increases (Fig. 3.5b). If the number of PSUs is increased in proportion to the increase in chlorophyll, then the maximum photosynthesis per unit area will also be proportional to the chlorophyll concentration (Fig. 3.5c), but this means that, on a chlorophyll basis, P_{max} will be constant (Fig. 3.5d).

Although very few marine plants have been examined in sufficient detail to permit firm conclusions to be drawn about the strategy of pigment alterations, the dinoflagellate *Glenodinium* provides a good example of a marine unicell in which PSU size appears to increase in response to decreasing irradiance (Fig. 3.6). The diatom *Phaeodactylum* shows a similar pattern of P vs I curves,[11] and so do a number of freshwater phytoplankton species ('*Chlorella*-type'[116]). In *Phaeodactylum*, the changes in both chlorophyll concentration and P_{max} per unit chlorophyll occur very rapidly when the cells are transferred to a new irradiance. A 2 – 3-fold decrease in chlorophyll a (and a corresponding increase in P_{max}) were apparent within 8 hours of a 17-fold increase in irradiance, although the changes took about 24 hours when the irradiance change was in the opposite direction.[11] Changes, such as these, in response to alterations in growth irradiance have often been interpreted as 'intensity adaptation' of phytoplankton photosynthesis,[237] since cells grown in low light are apparently able to make more 'efficient' use of low irradiances in photosynthesis (only on a per cell basis, see Fig. 3.5). However, it is often misleading to consider photosynthesis in isolation from the other metabolic activities of the cells, and the full biological significance of these changes in P vs I curves can only be appreciated in relation to the balance between nutrient uptake, photosynthesis, storage and growth, which is discussed in Chapter 4.

Large numbers of P vs I curves have been measured for natural

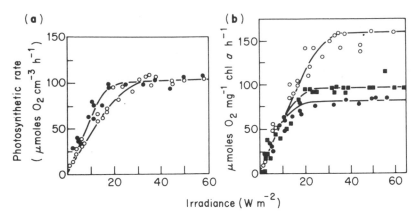

Fig. 3.6 P vs I curves for the dinoflagellate *Glenodinium*.[193] **(a)** Photosynthesis expressed per cm^3 of packed cells; **(b)** photosynthesis expressed per mg chlorophyll *a*. Growth irradiances: 2.5 W m^{-2} (●); 6.5 W m^{-2} (■); 25 W m^{-2} (○).

phytoplankton populations sampled from different depths, in different seasons and at different sites. Since it is impossible to separate phytoplankton cells from zooplankton and detritus, photosynthesis measurements can be expressed only on the basis of chlorophyll, and a full physiological analysis is not possible. The interpretation of these results is also complicated by the variations in species composition and in environmental factors that are liable to occur between samples. Thus, a decrease in P_{max} per unit chlorophyll between depths of 1 and 5 m, or between August and October, for example, cannot be interpreted simply — by analogy with studies in culture — as an increase in PSU size (see Fig. 3.5b), although it may mean that the second population contains a larger proportion of species with large PSUs. A particularly comprehensive study of P vs I curves of natural phytoplankton populations in the coastal waters off Nova Scotia showed that the initial slope and P_{max} were higher at 1 m than at 5 and 10 m, and that these parameters were also higher at all depths for two months during the summer but were fairly constant for the rest of the year.[191] The values for the initial slope correlated with the irradiance over the previous three days — a result which could have been due to an increase in the carotenoid:chlorophyll *a* ratio in response to high light (see p. 57) — but the changes in P_{max} were correlated with temperature and not with irradiance. The latter result suggests that the increase in P_{max} was due to higher enzyme activity, and not to changes in PSU size. This study shows that photosynthesis in the sea does not always conform to the pattern of laboratory experiments, or fit in with physiological theories, and it also emphasizes the extreme variability of phytoplankton samples and their activity. Conclusions that are based on anything less massive than the sampling programme carried through here (188 experiments!) should be treated rather cautiously.

TEMPERATURE AND MARINE PHOTOSYNTHESIS

The hypothesis that photosynthesis consists of a light and a dark reaction (see p. 44) predicts that temperature should have little effect on the rate of light-limited photosynthesis, but that the light-saturated rate should approximately double for every rise of $10\,^{\circ}C$ (i.e. $Q_{10} = 2.0$). These predictions have been supported by most laboratory studies of marine plants and, in the sea, phytoplankton P_{max} was found to be correlated with temperature, whereas the initial slope was correlated with irradiance (see above). Several phytoplankton species have also been cultured at different temperatures, and the P_{max} of cells from each of these growth temperatures has been measured at all of the other temperatures. Initial results with the diatom *Skeletonema*[115] and the green flagellate *Dunaliella*[163] indicated that cells grown at low temperatures had a higher protein content and showed higher activity of carbon-fixing enzymes than cells grown at higher temperatures, with the result that P_{max} at the growth temperature showed little change over the range 7 to $20\,^{\circ}C$. More recent results with both *Dunaliella* and two species of marine diatoms (*Phaeodactylum* and *Nitzschia*) have demonstrated that this evidence for 'temperature adaptation' may have been an artefact of the culture conditions.[164] When phytoplankton cells are grown in batch cultures, the nutrient status of the cells changes very rapidly, and cultures at high temperatures tend to reach a peak and start to decline before the cultures at low temperatures have reached a maximum. The timing of comparative experiments is, therefore, critical. If cultures grown at different temperatures are compared at their peaks of photosynthetic activity, P_{max} at low temperatures is found to be much the same regardless of growth temperature, but cultures grown at low temperatures are less able to assimilate CO_2 at high temperature than are cultures grown at higher temperatures.

Thus, phytoplankton cells seem to be capable of phenotypic adjustments to their photosynthetic apparatus (presumably, in this case, to the enzyme complement) in response to the prevailing temperature, but it is doubtful whether cells grown at low temperatures will ever exhibit greater photosynthetic rates than cells grown at high temperatures. Once again, it has been necessary to revise some of the earlier ideas about the physiological adaptation of plants to their environment, largely because photosynthesis had previously been considered in isolation from nutrient uptake and growth (compare p. 59). This topic of physiological adaptation is discussed further in Chapter 4 (p. 77).

SALINITY AND MARINE PHOTOSYNTHESIS

Although high salinity is the diagnostic feature of the marine environment, it appears to have very little effect on the photosynthesis and growth of marine plants. Salinity variations in the open sea are relatively small (see p. 20), and so the few studies of the effects of salinity on photosynthesis have mainly involved benthic plants from the intertidal zone. Most investigators have obtained reduced salinities by simply diluting sea water with distilled water, but a few have recognized that this is not a very scientific procedure, and neither is it typical of many natural situations. As far as photosynthesis is concerned, the total CO_2 content of the sea water is probably of greater importance than the absolute concentration of salts or the osmotic pressure. This hypothesis was elegantly tested by Ogata and Matsui[182] by exposing three intertidal Japanese seaweeds (*Ulva, Porphyra, Gelidium*) and the seagrass *Zostera* to salinities ranging from zero to twice that of sea water in each of three media: natural sea water, artificial sea water containing 5 mM $NaHCO_3$, and artificial sea water without $NaHCO_3$. The ranges of pH and ΣCO_2 over the salinity range tested are shown below:

	Natural sea water	Artificial sea water With $NaHCO_3$	Without $NaHCO_3$
pH range	6.0 − 9.1	7.9 − 8.5	5.6 − 6.6
ΣCO_2 range (mM)	0.08 − 3.92	4.82 − 5.20	0.63 − 1.00

The photosynthesis of *Porphyra* at a saturating irradiance over the full salinity range in each medium is shown in Fig. 3.7, and these results are typical of those obtained with all of the species. It is clear that the deleterious

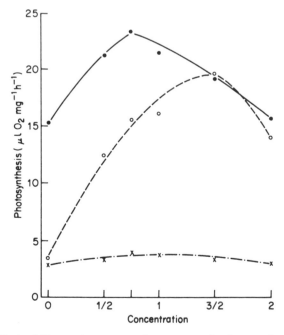

Fig. 3.7 Effects of dilute and concentrated sea water on the photosynthesis of *Porphyra tenera*.[182] Natural sea water (○); artificial sea water with $NaHCO_3$ (●); artificial sea water without $NaHCO_3$ (×).

effects of diluting natural sea water are largely relieved if the ΣCO_2 is maintained at a high level, as in the artificial sea water containing $NaHCO_3$. The poor photosynthesis in the artificial sea water lacking $NaHCO_3$ also shows that the salt content — or the osmotic pressure — of sea water has no positive benefit to marine plants in the absence of available carbon.

CARBON FIXATION IN MARINE PLANTS

Sources of inorganic carbon for marine photosynthesis

The light reaction of photosynthesis achieves energy fixation for plants, and the dark reaction achieves carbon fixation. Terrestrial plants are entirely dependent on atmospheric CO_2 as a source of inorganic carbon but, in addition to the dissolved gas, aquatic plants are exposed to two other forms of dissolved inorganic carbon: bicarbonate and carbonate ions (see p. 26–30). It is technically rather difficult, however, to tell whether a specific plant is able to utilize any form of carbon other than CO_2, and even more difficult to show which form is being taken up under given conditions. The ability to utilize bicarbonate ions (HCO_3^-) has been demonstrated in a few marine plants, usually by comparing the rate of photosynthesis at low pH, when the P_{CO_2} is high, with that at high pH, when HCO_3^- is the dominant form of inorganic carbon (see Fig. 2.9). Bicarbonate ion utilization is indicated if photosynthesis at high pH does not decrease as rapidly as the concentraton of free CO_2. For example, in a closed system, *Ulva* has been shown to photosynthesize steadily in spite of a sharp decrease in P_{CO_2} (Fig. 3.8a), whereas the photosynthetic rate of the brown alga *Desmarestia* followed the P_{CO_2} curve fairly closely (Fig. 3.8b). These experiments inevitably involve

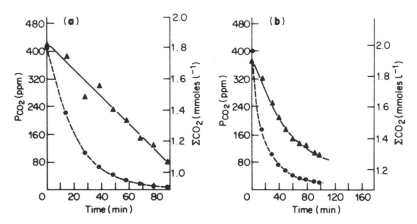

Fig. 3.8 Change of total inorganic carbon (ΣCO_2, ▲) and partial pressure of CO_2 (P_{CO_2}, ●) with time in a closed vessel containing **(a)** *Ulva* sp., and **(b)** *Desmarestia munda*.[110] The slope of the ΣCO_2 curve is a measure of the rate of net photosynthesis.

exposing plants to pH conditions that they would rarely, if ever, experience in the sea, but pH itself does not seem to be a critical factor. Once the photosynthetic rate of *Ulva* has fallen because of a reduction in ΣCO_2, it cannot be restored by adding acid, which decreases pH and increases P_{CO_2}, but only by adding $NaHCO_3$, which increases ΣCO_2 as well as P_{CO_2} (Fig. 3.9).

Similar experiments have shown that the brown alga *Sargassum* and the red alga *Iridaea* have the ability to take up HCO_3^-,[110] whereas the seagrass *Thalassia* does not,[13] and there is similar, but less complete, evidence that the laminarians *Alaria*, *Nereocystis* and *Costaria* may take up HCO_3^-, but that the green alga *Enteromorpha* and the red alga *Porphyra* do not.[110] Whether these species derive any ecological advantage — or suffer any disadvantage — as a result is still an open question, and must remain so at present, since it is not possible to demonstrate HCO_3^--uptake by macroscopic plants under conditions of pH and carbon concentration which are typical of sea water. A technique has recently been developed for freshwater planktonic algae, however, which may help to resolve the problem for

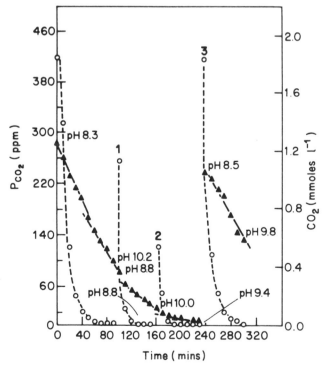

Fig. 3.9   Change of ΣCO_2 (▲) and P_{CO_2} (○) with time in a closed vessel containing *Ulva fenestrata*.[110] At **1** and **2**, HCl was added to restore pH and P_{CO_2} to original values; at **3**, $NaHCO_3$ was added to increase ΣCO_2 as well as P_{CO_2}. Photosynthetic rate (i.e. slope of ΣCO_2 curve) is related to ΣCO_2 and is independent of pH and P_{CO_2}.

marine phytoplankton. This technique depends on the fact that the liberation of free CO_2 from HCO_3^- is a relatively slow reaction, taking about 80 seconds to reach equilibrium.[233] If, therefore, cells which are able to utilize only CO_2 are provided with labelled carbon in the form of $H^{14}CO_3^-$, there should be a time-delay before ^{14}C appears in the cells, because the $H^{14}CO_3^-$ has to be converted to CO_2 before it can be assimilated. Time-courses of ^{14}C-uptake over very short time periods (up to 80 s) have shown that ^{14}C begins to appear in the cells of several freshwater green algae as soon as $H^{14}CO_3^-$ is supplied,[76] and comparable experiments with marine unicells (the diatom *Phaeodactylum* and the green flagellate *Dunaliella*) have shown the same result.[12, 166] These experiments confirm that at least some phytoplankton species can utilize HCO_3, but far more species need to be examined before bicarbonate uptake can be said to be widespread among marine plants, and there is no reason as yet to revise the conclusion reached earlier (p. 30) that bicarbonate uptake will rarely confer a significant ecological advantage in the sea.

Biochemistry of carbon fixation

It is slightly easier to find out *how* the carbon is being fixed in the cell than it is to discover which form of carbon is being fixed, and information is now available on the biochemistry of carbon fixation in several species from each of the major groups of marine plants. In terrestrial plants, primary carbon fixation occurs via ribulose bisphosphate carboxylase (RuBP-Case; C-3 plants) or via phospho-enol-pyruvate carboxylase (PEP-Case; C-4 plants), but all of the macroscopic marine plants appear to be typical C-3 plants. In red, brown and green seaweeds[128] and in seagrasses,[13] phosphoglyceric acid is the first labelled product of photosynthesis, and RuBP-Case is the dominant carboxylating enzyme. PEP-Case is detectable in a few seaweeds, but its contribution to carbon fixation seems to be negligible.[128] The situation in marine phytoplankton is rather less clear-cut because reports of high PEP-Case activity in some diatoms and flagellates[12] have been contradicted by other results.[127] It appears that C-4 photosynthesis can occur in marine diatoms, but is significant only when cell densities are high and when CO_2 depletion is rapid. The ratio of RuBP-Case to PEP-Case is high in cultures which are in the logarithmic phase of growth, when photosynthesis per cell is maximal, but decreases in stationary-phase cultures when cell numbers are high.[83, 166]

Thus, the biochemical evidence suggests that most marine plants in most conditions follow the C-3 pathway of carbon fixation, and the stimulation of net photosynthesis by reduced oxygen concentrations in several tropical marine plants[245] supports this conclusion by indicating the occurrence of photorespiration. Certain other physiological characteristics of some marine plants, however, resemble those of terrestrial C-4 plants. The available measurements of CO_2-compensation points indicate extremely low values (< 10 ppm[45, 137]), more akin to C-4 than to C-3 plants, and the ratio between the stable isotopes of carbon ($^{13}C : ^{12}C$) in many marine plants is also closer to that found in C-4 plants.[13, 245] The apparent paradox of plants possessing the biochemical characteristics of C-3 plants and the physiological characteristics of C-4 plants may be due to the high resistance to CO_2-diffusion imposed by the cell membrane, so that the internal and external supplies of inorganic carbon are not in equilibrium.[13] Taking up bicarbonate may also provide a mechanism for concentrating CO_2 inside the cells of aquatic plants, and so reducing the apparent CO_2-compensation point.[16]

A further peculiarity of carbon fixation in marine plants is the apparently high activity of phospho-enol-pyruvate carboxykinase (PEP-CKase) in numerous marine algae, and especially in some diatoms[7, 127] and brown seaweeds.[128, 130] In higher plants, this enzyme catalyses the removal of CO_2 from oxalacetate to form PEP, but preparations of the enzyme from diatoms have substantially different properties, and act as carboxylases, combining CO_2 with PEP and forming ATP in the process.[7] This pathway of carbon fixation may enable marine plants to utilize HCO_3^- in the biosynthesis of metabolites with four carbon atoms, such as malate and aspartate, but the process is still dependent on RuBP-Case, since the source of the PEP is thought to be phosphoglyceraldehyde formed in the light. Enzyme preparations from brown algae show similar properties to the enzyme from diatoms, although they have yet to be fully purified and definitely identified as PEP-CKase. The activity attributed to PEP-CKase in brown algae varies markedly within a single thallus, and is strongly correlated with the rate of dark fixation of CO_2, which may be up to 30% of P_{max} in young tissues (e.g. meristematic regions of *Laminaria* fronds[130]). It has been suggested that PEP-CKase is the enzyme responsible for these high rates of dark CO_2-fixation, but it is not clear how the PEP would be regenerated in darkness. The detection of this additional carboxylating enzyme in several distinct groups of marine plants suggests that its occurrence is related to the marine habit, but the full ecological and physiological significance of PEP-CKase activity has yet to be established.

4

Growth and Productivity of Marine Plants

The growth of natural populations of both animals and plants is extremely difficult to measure under natural conditions. For animals, there is no alternative but to estimate the growth rate — for one species at a time — from changes in numbers and weights of individuals in the field, allowing for losses resulting from predation and death. For photosynthetic plants, however, it is possible — and very much easier — to estimate the overall productivity of all of the species in an ecosystem by measuring the uptake or release of inorganic chemicals (e.g. CO_2, O_2, $NO_3^- + NH_4^+$) in the environment, and most attention has been concentrated on this approach. These rates of exchange of inorganic chemicals between plants and their environment are correlated with the rates of individual processes that contribute to growth, such as photosynthesis and ion uptake, rather than with growth rate itself, and direct studies of plant growth in the sea have been largely restricted to a few species, mainly kelps, which are particularly convenient for such work (see p. 83). It is, therefore, important to examine the relationships between photosynthesis, nutrient uptake and growth, as revealed by laboratory experiments, before attempting to interpret the seasonal and geographical variations of plant biomass and primary productivity in the sea.

NUTRIENT UPTAKE

Requirements for growth

The algae are physiologically more diverse than vascular plants, and this diversity is reflected in their nutrient requirements. As a group, they require a larger range of elements in an inorganic form than higher plants; many species are able to obtain some essential elements (especially N and P) from organic compounds: other species have absolute requirements for certain organic compounds, such as vitamins.[183] Another difference from vascular plants is that, in some algae, it is possible to replace some of the 'essential' elements by chemically related elements (e.g. K by Na, Ca by Sr). Natural sea water contains such high concentrations of many of the elements that are regarded as 'macronutrients' for plant growth (e.g. K, S, Ca, Mg; see Table 2.2) that it is very difficult to demonstrate that they are essential for marine plants. An artificial sea water medium that was completely free of potassium or sulphur, for example, would be so artificial that it would be dangerous to attribute poor growth in such a medium simply to the absence of the one element that had been omitted. However, the metabolic functions of these macronutrient elements in higher plants are so fundamental for plant growth that it is generally assumed that they are essential for marine plants as well. Sulphur has an additional role in many seaweeds as a component of sulphated polysaccharides (e.g. fucoidan, carrageenans) which are important in attaching plants to their substrate (see p. 98) as well as being major commercial products (see p. 158). Nitrogen and phosphorus are macronutrients whose concentration in sea water is relatively low and variable (see Table 2.4) and, therefore, much of this chapter is concerned with the influence that these two elements exert on the growth of marine plants.

Most of the elements that have been identified as essential 'micronutrients' for vascular plants (e.g. Fe, Mn, Cu, Zn, Mo) are required in such low concentrations that, again, it is difficult to demonstrate that marine plants have an absolute requirement for them. None of these metals is abundant in normal sea water (see Table 2.4), and the absolute concentrations determined by chemical analysis are not necessarily the concentrations that are biologically effective. The uptake of trace metals is often facilitated by organic compounds in sea water which act as natural chelators, and the presence of these compounds may be as important as the presence of the metals themselves.[207] It is clearly preferable to test for nutrient requirements in completely defined, artificial media, rather than in media based on natural sea water, but here the problem is no easier, since the contaminants of the major salts (even if 'Analar' grade reagents are used) contribute sufficient of the micronutrients to satisfy the requirements of most marine plants. Marine algae are, therefore, assumed to require small amounts of Fe, Mn, Cu, Zn and Mo, because of the metabolic importance of these elements in higher plants, but there is little direct evidence that they are essential for marine species. On the other hand, high concentrations of copper and zinc, along with other heavy metals, are well known as inhibitors of algal growth, and are used to prevent 'fouling' of ships' bottoms, piers and slipways (see p. 173). Another essential micronutrient for higher plant growth, boron, has been shown to be essential for some marine algae but not for others. Boron was found to be required for the growth of many diatoms, but not for several marine flagellates,[133] and the failure to enrich natural sea water with boron prevented the growth of *Fucus* embryos in culture.[157] The latter observation suggests that *Fucus* has a high boron requirement, since this element is one of the more abundant in sea water (see Table 2.2). The metabolic role of boron in these plants has not yet been identified, and neither has that of iodine and bromine, which appear to be essential for some brown and red seaweeds (e.g. *Polysiphonia*, Br and I;[79, 80] *Ectocarpus*, I;[266]

Fucus, Br[157]), although they are not required by higher plants. The role of silicon in the cell walls of diatoms and the skeletons of silicoflagellates is quite clear, however, and it is not surprising to find that most diatoms have an absolute requirement for this element. Other marine algae may contain substantial amounts of silicon, but do not require it for growth (e.g. *Fucus*[157]).

Absolute requirements for vitamins, or the ability to utilize organic sources of nitrogen and phosphorus, can be demonstrated only in bacteria-free cultures of marine algae, since, if bacteria are present, they may synthesize the vitamins that the alga requires, or they may release inorganic N and P from the organic sources. So few macroscopic seaweeds have been successfully cultured in the absence of bacteria (see p. 106) that the information available on these aspects of algal nutrition is strongly biassed in favour of unicellular forms. Nearly all dinoflagellates, cryptomonads and chrysophytan flagellates require either vitamin B_{12} or thiamine, but only about half of the diatoms and 40% of the green algae tested require these vitamins for growth.[198] Ten species of red algae (all of the multicellular forms tested) have been shown to require vitamin B_{12}, but the unicellular red alga *Porphyridium* and nine out of ten brown algae investigated had no requirements for any vitamins. These requirements — or lack of them — are often difficult to demonstrate, even after the alga has been separated from bacteria, because very low concentrations (< 1 ng l^{-1}) are often sufficient. Such requirements could be ecologically important, however, because the average concentration of vitamin B_{12} in sea water varies from $2-3$ ng l^{-1} in coastal waters to as low as 0.1 ng l^{-1} in the open ocean, and the seasonal variations in concentration are similar to those of nitrogen and phosphorus (see p. 32).

A wide range of organic nitrogen compounds, including urea, various amino acids and amides, and the purine hypoxanthine, have recently been shown to support the growth of most of the planktonic marine algae tested, which included representatives from all of the algal divisions except the Phaeophyta.[6, 247] The red algae were represented only by *Porphyridium* (which, judging from its vitamin requirements, may not be typical of the group) and this species was unable to utilize even urea, but the only brown alga examined to date (*Ectocarpus*) utilized several different types of organic N-source.[176] Antia *et al.*[5] have placed particular stress on the utilization of hypoxanthine and guanine, since these purines have been identified as a common excretory product of smaller zooplankton organisms (e.g. ciliates). It has long been thought that complex excretory products of this type would have to be broken down by bacteria before the nitrogen would be available for re-assimilation by phytoplankton, but several lines of evidence now suggest that nutrients can be recycled more rapidly than this in the marine planktonic system (see p. 74).

Mechanism of ion uptake

Although sea water contains higher concentrations of most of the elements required for plant growth than fresh water or the soil solutions available to most terrestrial plants, this does not mean that marine plants can rely on passive diffusion for their nutrient supplies. Ion uptake by plants in the sea — as in all other environments — is an active process, requiring energy derived from cellular metabolism. The rate of uptake of sulphate by excised apices of *Fucus,* for example, was halved in darkness and in the presence of photosynthetic inhibitors, while inhibitors of respiration suppressed uptake completely.[44] The individual cells of some species of marine algae (the so-called 'giant-celled' forms, such as the red alga *Griffithsia*, and the green algae *Chaetomorpha* and *Acetabularia*) are so large that it is possible to measure the cytoplasmic and vacuolar concentrations of major ions separately, and to estimate the electrical potential across the plasmalemma and the tonoplast. Work with such species has shown that, whereas most ions are taken up actively from sea water, against an electrochemical gradient, sodium is actively pumped out of the cells.[150] Since sodium is not an essential element for the growth of most plants but is present in very high concentrations in sea water, the ability of these (and possibly all) marine plants to expel sodium actively from the cells could possibly be seen as a clear ecological adaptation to life in the sea. However, other giant-celled algae that are entirely freshwater forms (e.g. *Chara, Nitella, Hydrodictyon*) exhibit the same active efflux of sodium from the cytoplasm to the external medium,[150] and so the fundamental difference between marine and freshwater plants (see Chapter 1) does not seem to lie in their ion uptake mechanisms.

Further evidence for active mechanisms of ionic exchange between marine plants and the sea has been obtained through studies of the kinetics of ion uptake by marine algae. The rate of uptake of an ion often increases in direct proportion to the external concentration of that ion at low concentrations, but gradually levels off and becomes saturated at higher concentrations (Fig. 4.1). The typical hyperbolic form of this graph means that uptake cannot occur simply by passive diffusion, since this would result in an indefinite continuation of the initial linear relationship, but it is consistent with the hypothesis that a 'carrier' actively transports the ions across the cell membrane. On the basis of this hypothesis, a theoretical equation can be derived, which relates the rate of uptake (V) of an ion to its concentration (S) in the external medium, and which fits the hyperbolic pattern of the observed results:

$$V = \frac{V_{max} \cdot S}{K_t + S} \qquad (4.1)$$

where V_{max} = maximum rate of uptake under the conditions of the experiment,

and K_t = half-saturation constant for uptake (or transport; i.e. the external concentration at which $V = \frac{1}{2} V_{max}$).

The constants in the equation, V_{max} and K_t, can be estimated from experimental results of the type shown in Fig. 4.1, and K_t-values are available for numerous species of unicellular marine algae (mainly for N-, P- and Si-

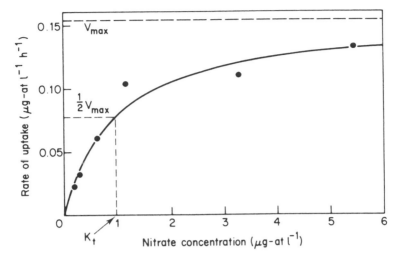

Fig. 4.1 Rate of uptake of nitrate by a natural phytoplankton population from the Pacific Ocean, incubated *in situ* with different concentrations of added nitrate.[149] The curve was calculated from equation 4.1, with $V_{max} = 0.156 \,\mu\text{g-at I}^{-1}\,\text{h}^{-1}$ and $K_t = 0.98 \,\mu\text{g-at I}^{-1}$.

uptake[72]), for natural populations of marine phytoplankton (e.g. Fig. 4.1), and for sulphate uptake by *Fucus*.[44] The similarity between equation 4.1 and the familiar Michaelis-Menten equation of enzyme kinetics has meant that low K_t-values have often been interpreted — by analogy with the Michaelis constant, K_m — as an indication of high affinity between the cells and specific ions. However, the theoretical derivation of equation 4.1 is somewhat different from that of the Michaelis-Menten equation,[175] and the K_t-value for a species does not necessarily reflect the affinity of that species for the ion to which it refers. For example, in a group of species with identical uptake rates for a given ion at low concentrations, the K_t-value for each species will be proportional to the maximum uptake rate (V_{max}) of that species, and not to its ability to take up the ion at low concentrations. The ecological interpretation of ion uptake kinetics should, therefore, be approached with considerable caution.

Nutrient supply and growth

It is not only the rate of uptake of nutrients that shows a hyperbolic relationship to external nutrient concentration. If a planktonic alga is grown under conditions in which only one nutrient is limiting, the growth rate of the alga often shows a similar hyperbolic relationship to the concentration of the limiting nutrient in the external medium (Fig. 4.2), and this relationship can be described by an equation (the Monod equation) which is analogous to the equation for ion uptake:

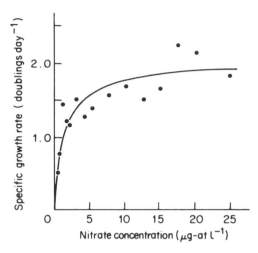

Fig. 4.2 Specific growth rate of nitrogen-depleted cells of *Asterionella japonica* at various concentrations of external nitrate.[73] The curve was calculated from equation 4.2, with $\mu_{max} = 1.97$ doublings day^{-1} and $K_g = 1.5\,\mu$g-at l^{-1}.

$$\mu = \frac{\mu_{max} \cdot S}{K_g + S}, \qquad (4.2)$$

where μ = specific growth rate (in doublings day^{-1}) at external nutrient concentration S, and μ_{max} is the maximum growth rate under the conditions of the experiment (i.e. when the nutrient is no longer limiting growth rate);

and K_g = half-saturation constant for growth.

Unlike equation 4.1 and the Michaelis-Menten equation, the Monod equation is purely empirical. It provides an adequate fit to the experimental results, but it has not been developed from any basic hypothesis about the physiological mechanisms that relate growth rate to nutrient supply. The value of K_g cannot, therefore, be used as an indicator of the ability of a species to grow at low nutrient concentrations. However, K_g-values do provide an indication of the range of concentrations over which the growth of a species is liable to be limited by a particular nutrient.

A comparison of the concentrations of inorganic nitrogen and phosphorus in sea water with experimentally determined K_g-values for common phytoplankton species suggests that the growth of natural phytoplankton populations is often limited by the availability of these elements. Indeed, in oceanic waters, the concentration of nitrogen is frequently below the limits of analytical detection, so that oceanic phytoplankton would be expected to show very low growth rates. Nevertheless, such waters contain phytoplankton cells which appear to be healthy, and which photosynthesize actively, although their numbers may be low in comparison with those of coastal waters. Another paradox is that the average atomic ratio of C:N:P in marine phytoplankton is 106:16:1 (see p. 31), but this ratio is obtained in culture only if the cells are not limited by nitrogen and phosphorus. At low levels of external phosphorus, for example, the flagellate *Monochrysis* exhibits a C:P ratio of over 1000:1, but this ratio decreases as the phosphorus supply and the growth rate increase, and approaches 106:1 only when the growth rate approaches μ_{max} and the cells become phosphorus-saturated (Fig. 4.3). These results suggest that most natural phytoplankton populations, with low C:P and C:N ratios, are not nutrient-limited, even though the nutrient content of the sea water often seems to be below that required to saturate growth. The explanation of this apparent conflict between laboratory results and field observations may lie in recent experiments on short-term ion uptake, and in the utilization of organic sources of nitrogen.

If the marine diatom *Skeletonema* is grown under severe nitrogen-limitation for a period and the medium is then enriched with a single addition of ammonium, there is an initial phase of rapid uptake, lasting for up to 20 min (V_s, Fig. 4.4), followed by a period of 2 – 3 hours at a lower, constant rate of uptake (V_i). Only when the ammonium in the medium has been substantially reduced does the uptake rate begin to show the expected relationship to external ammonium concentration (V_e, Fig. 4.4). This pattern of changing uptake rates suggests that there is an intracellular pool of inorganic ammonium, which is rapidly filled by the initial surge of uptake, and that the rate of uptake during the second phase (V_i) is controlled by the rate at which ammonium is removed from the pool and incorporated into

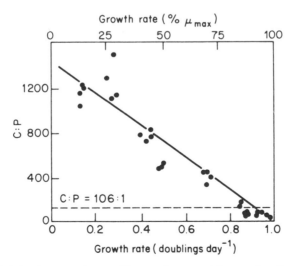

Fig. 4.3 Relation between the specific growth rate and the atomic ratio of carbon to phosphorus (C:P) in *Monochrysis lutheri* under phosphorus limitation in continuous culture (μ_{max} = 0.95 doublings day^{-1}).[84] The broken line represents the approximate ratio found in natural phytoplankton populations (i.e. 106:1). (Reprinted with permission. © 1979 Macmillan Journals Limited.)

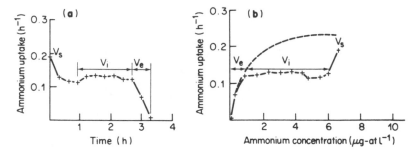

Fig. 4.4 Ammonium uptake (expressed as doublings of the cell nitrogen content per hour) by nitrogen-starved cells of *Skeletonema costatum*.[42] **(a)** Rates of uptake at different times after a single addition of ammonium, raising the concentration from <0.5 to >7 μg-at l^{-1}; **(b)** the same results expressed as a function of the residual ammonium concentration. The broken line in **(b)** represents an extrapolation of phase V_e.

cellular materials, rather than by the external concentration of ammonium. A similar initial surge of ammonium uptake has been observed in another marine diatom, *Thalassiosira*. Nitrogen-limited cells, growing at a rate of 0.26 doublings day^{-1}, were completely starved of nitrogen for 2 hours and then exposed to 16 μg-at NH$^+_4$ l^{-1}. During the first 5 minutes of incubation, the ammonium uptake (expressed as doublings of the cell nitrogen content per day) was 8.8 — i.e. 33.9 times faster than the growth rate[154]. This result was not thought to mean that this species could possibly sustain a growth rate of 8.8 doublings day^{-1} (μ_{max} in this experiment was 1.01 doublings day^{-1}), but it does suggest that the cells could be exposed to ammonium for as little as 42 minutes per day (i.e. 1/33.9 × 24 hours) and still obtain sufficient nitrogen to support their current growth rate. The lower the rate of growth, and the longer the period of complete nitrogen starvation, the more rapidly the cells took up the ammonium when it was provided (Table 4.1).

Individual cells in the phytoplankton do not, therefore, need to be exposed continuously to all of their essential nutrients, because they are able to take up these nutrients rapidly from small, transient patches of high concentration. Such patches could result from the decomposition of other algal cells, or the excretion of planktonic animals. Some oceanic copepods, for example, have been found to excrete 2.6 × 10^{-7} μg-at N per animal in 5 seconds,[154] and this means that, within the volume displaced by the animal (about 5 × 10^{-8} litres), the available nitrogen could increase by about 5 μg-at l^{-1}. Since the individual cells of most oceanic phytoplankton are substantially smaller than such copepods, this represents a rich supply of nitrogen, provided that the nutrient is not rapidly dispersed by water movement.

Thus, the discrepancies between laboratory measurements and field observations may arise partly from the use of inappropriate scales for the measurements. Ion uptake rates have generally been measured over periods of hours, whereas phytoplankton cells appear to respond to changing concentrations within minutes. Similarly, nutrient concentrations in the sea have been measured in (and averaged over) litres, whereas the typical

Table 4.1 Maximum daily rate of ammonium uptake (expressed as doublings of the cell nitrogen content) by an oceanic clone of the diatom *Thalassiosira pseudonana*, measured at three growth rates after various periods of complete nitrogen starvation.[154] The maximum growth rate (μ_{max}) of the clone was 1.01 doublings day^{-1}.

Length of nitrogen starvation (minutes)	Specific growth rate (μ) (doublings day^{-1})		
	0.26	0.60	0.82
0	4.00	2.40	0.90
20	6.68	4.02	0.90
120	8.81	4.92	1.15

volumes of nutrient patches and phytoplankton cells are some 15 orders of magnitude smaller (N.B. 1 litre $= 10^{15} \mu m^3$). Another factor contributing to these discrepancies may have been that most measurements of nitrogen and phosphorus availability in the sea have concentrated on the inorganic forms of each element (i.e. NO_3-N or NH_4-N and PO_4-P). Recent measurements of total dissolved nitrogen and total dissolved phosphorus in the English Channel have indicated, however, that when the concentrations of NO_3-N and PO_4-P fall during periods of rapid phytoplankton growth (see Fig. 2.10), the concentrations of dissolved organic compounds of nitrogen and phosphorus rise, so that there is relatively little change in the total dissolved N and P in the water throughout the year.[26] Since the majority of unicellular marine algae seem to be able to utilize several organic compounds as nitrogen sources (see p. 69), including fairly complex compounds such as purines, these results provide further support for the hypothesis that the growth rates of marine phytoplankton in the sea are not necessarily limited by nutrient supplies, but may be sustained by rapid recycling of essential elements within the photic zone. The low standing crop of phytoplankton in oceanic areas and in the surface layers of seasonally stratified waters may, therefore, be more closely related to the grazing pressure imposed by zooplankton than to nutrient availability (see p. 82).

GROWTH, STORAGE AND EXCRETION IN RELATION TO IRRADIANCE AND NUTRIENT SUPPLY

The other major process contributing to plant growth is photosynthesis, and the relationship between growth and irradiance resembles that between photosynthesis and irradiance (compare Fig. 4.5 with Fig. 3.6). Unfortunately, there have been few direct comparisons between growth versus irradiance curves and P vs I curves for the same species grown under similar conditions, but the results that are available show that growth is generally light-saturated at lower irradiances than photosynthesis (Table 4.2). The most probable explanation for this observation is that the supply of inorganic nutrients, which will have little direct effect on the rate of photosynthesis, has become limiting for growth before the rate of the dark reaction has become limiting for photosynthesis. According to this hypothesis, the

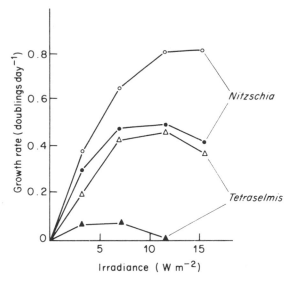

Fig. 4.5 Specific growth rates of the diatom *Nitzschia* (circles) and the flagellate *Tetraselmis* (triangles) as a function of irradiance at high and low nutrient concentrations.[151] Open symbols: 10 mg-at N and 0.59 mg-at P I^{-1}; filled symbols: 8.9 μg-at N and 0.42 μg-at P I^{-1}.

Table 4.2 Saturating irradiances for photosynthesis and for growth in various marine algae.

Algal group	Species	Saturating irradiance (W m^{-2}) Photosynthesis	Growth
Chlorophyta	*Acetabularia crenulata*	12 – 16	3
Phaeophyta	*Fucus serratus*	100	30
	Laminaria saccharina	30	14
Rhodophyta	*Chondrus crispus*	36	19
Pyrrhophyta	*Glenodinium* sp.	20 – 30	10

References: *Acetabularia;*[94, 244] *Fucus, Laminaria, Chondrus;*[144] *Glenodinium*[193]

irradiance required to saturate growth should be decreased in low nutrient concentrations, and this is supported by the results in Fig. 4.5. At very high nutrient levels, however, growth and photosynthesis should become light-saturated at similar irradiances. This second prediction is supported by some observations on phytoplankton[108], but it is possible, particularly in macroscopic plants, that growth will always be saturated at lower irradiances than photosynthesis, because the maximum rate of nutrient uptake (V_{max} in equation 4.1), or the rate of transport of nutrients through the thallus, is insufficient to meet the nutrient demand created by the maximum rate of photosynthesis (P_{max}).

Whenever the growth of a plant is limited by the supply of inorganic nutrients, rather than by the rate of photosynthesis, the plant will be producing more photosynthate than can be used in growth, or is required in respiration. This excess photosynthate must either accumulate within the plant as storage polysaccharides (see Table 1.2 for the typical compounds in different algal groups), or be excreted into the external medium. There is a clear competitive advantage in storing excess carbohydrate, at least during the early stages of nutrient-limited growth, but the accumulation of carbon which has no metabolic or structural role inside the cell must produce some changes in the cell's metabolism. In most multicellular plants, of course, the excess photosynthate can be exported from the photosynthetic cells and stored in specialized storage tissues, but this is not possible in unicellular algae or in multicellular algae in which all cells are photosynthetic. As reserve carbon accumulates, the proportion of the total cell carbon available for synthetic machinery must fall, and either the biosynthetic or the photosynthetic apparatus (or both) of the cell must be reduced.

If the biosynthetic apparatus is reduced, the growth rate will decrease at any irradiance above that required to saturate growth. However, the growth rate does not normally decrease until substantially higher light levels are reached (e.g. Fig. 4.5). It is the content of chlorophyll (i.e. the photosynthetic apparatus) of many algae that decreases at high growth irradiances (see p. 55, Table 3.4). This response can now be interpreted as an adaptation to reduce the energy intake of the cells, and to prevent the excessive accumulation of reserves, while maintaining growth at the maximum rate permitted by the available nutrient supply. A theoretical model of the physiological adaptation of unicellular algae has recently been developed,[231] based on the hypothesis that the non-structural carbon of the cell can be divided into three components (biosynthetic apparatus, photosynthetic apparatus, and reserve carbon), and that the relative proportions of these components are adjusted so that, under any given environmental conditions, the growth rate is maximized, and all cell components increase at the same relative rate. The model successfully predicts the decrease in chlorophyll content and the increase in nucleic acids that is often observed in response to increased growth irradiance, as well as the decrease in both chlorophyll and nucleic acids and the increase in stored carbon that occur under conditions of nutrient deficiency (Fig. 4.6). Although the details of this model will probably have to be modified in the course of experimental validation, it seems to offer a more promising approach to physiological adaptation than considering photosynthesis in isolation from growth, and studying the influence of different environmental factors separately (see pp. 52, 59, 61).

A possible alternative to storing excess carbohydrate inside the cells is to excrete it into the external medium, although this will probably not occur until the cell's reserves are fairly high, and perhaps only when further increase would reduce the biosynthetic and photosynthetic apparatus below some minimum 'safety level' (see Fig. 4.6). It has long been known that, under certain conditions, phytoplankton cells release soluble organic compounds (often simple carbohydrate molecules, such as glycollic acid)

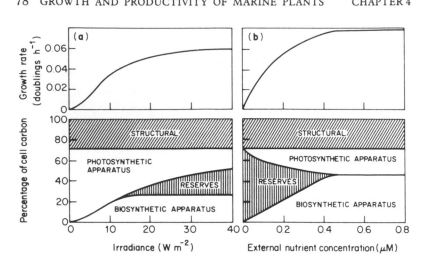

Fig. 4.6   Effects of irradiance (a) and nutrient supply (b) on the growth rate (in doublings h^{-1}) and the percentage of total cell carbon in each of four cellular components of a unicellular alga with a cell diameter of 20 μm at 25°C, as predicted by a theoretical model.[231] Nutrient supply in (a) was 0.25 μM, and irradiance in (b) was 40 W m^{-2}. All calculations were based on values given by Shuter.[231]

into the surrounding water, but there is little agreement about exactly why and when such excretion occurs, and some authors have questioned whether 'healthy cells do it'.[229] If excretion is interpreted as a means of avoiding excessive accumulation of reserve carbon under nutrient-limited growth conditions, it is possible to detect a pattern in most of the results from culture experiments and field investigations, and to resolve some of the apparent contradictions in the evidence. Most culture studies have shown that the percentage of photosynthetic carbon that is released as soluble organic compounds is small (3 – 6%) when cells are in the logarithmic phase of growth, regardless of irradiance, photosynthetic rate or cell density.[262] As a culture ages, however, the rate of excretion increases, and cells grown at low nutrient concentrations also show high excretion rates.[100] Higher excretion has usually been reported for natural populations in oligotrophic waters (up to 50% of photosynthetic carbon) than in eutrophic waters (0 – 10%),[207] and rates at the surface (where growth is probably light-saturated) and at depth (where cells are senescent) are higher than at intermediate depths.[15] There have been reports of low excretion rates in nutrient-poor waters, but these are not inconsistent with the hypothesis presented here, since the growth of the phytoplankton populations may have been light- or temperature-limited at the time of the measurements. The hypothesis that excretion provides a 'safety valve' for the release of excess photosynthate has still to be subjected to direct experimental test, but it has the attraction that excretion is no longer seen as an entirely wasteful process, but rather as one of several mechanisms available for regulating and maintaining the metabolic balance of cells under fluctuating environmental conditions.

TEMPERATURE AND GROWTH

The effects of temperature on plant growth are determined by the balance between photosynthesis, nutrient uptake and respiration. Of these three processes, the second two are primarily chemical phenomena, catalysed by enzymes, and these processes normally show a Q_{10} of about 2.0 (see p. 61) when their substrates are not limiting. The rate of photosynthesis, on the other hand, is often independent of temperature (i.e. Q_{10} = 1.0), particularly in warm conditions, because of limitation by irradiance or CO_2-supply. The most rapid growth of a species tends to occur, therefore, at temperatures higher than those at which photosynthetic rate is temperature-limited, but not so high that most of the photosynthate is used up in respiration. Such optimal temperatures for growth have been determined for many phyto-plankton species in culture, but these temperatures are frequently higher than those recorded in the sea when the species are most abundant. For example, *Coccolithus* and the diatom *Asterionella* both grew best at $20-25°C$ in culture, but reach their maximum abundance in the sea at $7-10°C$.[207] There are several possible explanations for these discrepancies between laboratory and field observations (e.g. competition with other species in the sea, or changes in temperature requirements at suboptimal nutrient levels — see p. 143), but they may be caused by long-term pheno-typic or genotypic changes in the respiratory mechanism, which increase the respiration rate at low temperatures and decrease the rate at high temperatures. Genotypic changes of this type are known in flowering plants (ecotypes of alpine species from high altitudes respire more rapidly at low temperatures than ecotypes from lower altitudes[123]), and measurements of the respiration of kelps in the sea have indicated that similar phenotypic changes occur with season in these plants.

The respiration rate per unit area of blade from *Laminaria hyperborea* in the North Sea was about 50% higher in August than in March (Fig. 4.7b) but, when expressed on a dry weight basis, the respiration rate increased sharply as sea temperatures rose from $4°C$ in March to $8°C$ in May, but then declined as the temperature increased further to $16°C$ in August (Fig. 4.7c). Thus, the respiration of unit dry weight of *Laminaria* blade was almost the same in March and August, in spite of the $12°C$ difference in temperature, but the respiration rate of thalli adapted to winter tempera-tures doubled in response to a temperature rise of only $4°C$. This very rapid rise of respiration with temperature in the spring probably also reflects the effects of temperature on growth rate. *Laminaria* grows very rapidly between January and June (see Fig. 4.9, p. 84), but low water temperatures must limit the growth rate during the first half of this period. These results also reveal marked seasonal changes in the dry weight per unit area of blade tissue, which are related to alterations in the balance between growth, photo-synthesis and storage in kelps at different times of year (see p. 85). Many marine plants show similar seasonal patterns of growth in their natural habitat, and these can now be examined in the light of our knowledge about the marine environment, and the physiology of the processes that contribute to growth.

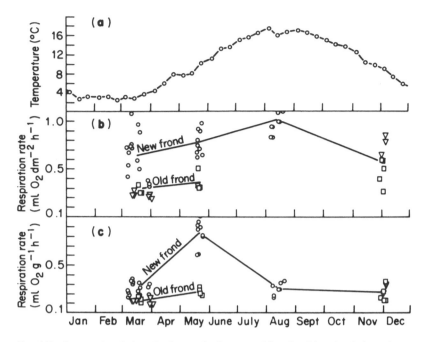

Fig. 4.7 Seasonal variations in the respiration rate of fronds of *Laminaria hyperborea* from the North Sea, measured in the laboratory at ambient sea water temperatures.[140] 'New frond' refers to tissue formed since the beginning of the year; 'old frond' to tissue that has persisted from the growing season of the previous year.

SEASONAL PATTERNS OF GROWTH IN THE SEA

Phytoplankton

Direct measurements of the growth rate of a natural phytoplankton population in natural conditions would involve enclosing a water sample, from which all grazing animals had been removed, in a transparent container, and suspending it in the sea for at least 3–4 days. The environment that the enclosed cells would experience over this period would be so different from that experienced by unenclosed cells (see p. 86) that this treatment — even assuming that the complete removal of the zooplankton was possible — would largely defeat the object of the experiment. For this reason, the seasonal pattern of phytoplankton growth in the sea must be deduced from regular measurements of the standing crop or the *in situ* photosynthetic rate of the phytoplankton. Such studies of the annual cycle of the phytoplankton in any region need to take account of the very patchy distribution of plankton — both horizontally and vertically — in natural waters, and the wide range of size among phytoplankton cells. Some sampling techniques fail to harvest the smaller cells, and estimates of the overall abundance of phytoplankton may be distorted by changes in the size distribution of the population. For example, 95% of the individual cells in a phytoplankton crop could consist of small flagellates of the nanoplankton (i.e. < 20 μm in diameter), but these cells might contribute less than 10% of the total chlorophyll content of the crop, whereas large diatoms (50 – 200 μm in diameter) could represent 80% of the total chlorophyll and less than 5% of the cell numbers. The results of individual surveys cannot be interpreted, therefore, without detailed reference to the sampling and estimation techniques, but certain features of the annual cycle are common to most surveys (Fig. 4.8) and it is these that are discussed here.

The change in phytoplankton crop between one sampling time and the next depends on the balance between three processes: cell growth, grazing by zooplankton, and cells sinking out of the sampling zone. During the winter months in temperate regions, zooplankton numbers are very low (Fig. 4.8b), and grazing can be assumed to be negligible. The low temperatures at the surface mean that there is little thermal stratification (see Fig. 2.8) and that the water column is well mixed, so that the net loss of cells by sinking will also be small. The very low phytoplankton crop must, therefore, be due to low growth rates, but these cannot be attributed simply to the wintry conditions at the surface. Light conditions will certainly be poor, because of both low irradiances and short daylengths, but the temperature of

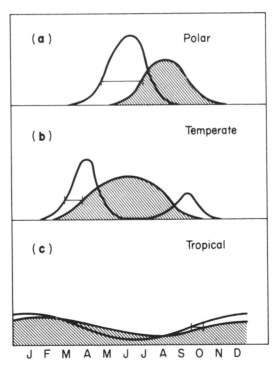

Fig. 4.8 Seasonal variations in the biomass of phytoplankton and of herbivorous zooplankton (hatched area) in different latitudes.[49] The horizontal bar indicates the delay period between the increases of phytoplankton and zooplankton.

most open seas in temperate latitudes does not fall below $4-5°C$ (Fig. 2.8), and this would be adequate for moderate growth of most temperate plants. The critical factor is the continual mixing of the water column, since this means that phytoplankton cells spend little time in the relatively shallow photic zone. Such cells will be able to grow only if the mean irradiance received over each 24-hour period exceeds the compensation irradiance for photosynthesis. If the surface irradiance and the attenuation coefficient for the water column (see p. 14) are known, it is possible to calculate the maximum depth of the mixed layer (the *critical depth*) that will support positive net growth of the phytoplankton. Provided that $D_{cr}. k > c. 2.5$,

$$\text{critical depth, } D_{cr} = \frac{I_o}{k.I_c}, \qquad (4.3)$$

where I_o (surface irradiance), I_c (compensation irradiance) and k (attenuation coefficient) all refer to photosynthetically active radiation ($400-700$ nm). When the mixed layer exceeds the critical depth, no net growth is possible because the mean irradiance is less than the compensation irradiance. This situation persists through most of the winter but, in the spring, the critical depth increases (largely due to increasing I_0) and, at the same time, the depth of the mixed layer decreases, since the surface waters begin to warm up and stratification sets in (see Fig. 2.8). The result is that the growth rate of the phytoplankton increases more rapidly than can be explained by the increases in surface irradiance and temperature and, in the absence of significant numbers of zooplankton, the crop rises dramatically, producing the so-called spring 'outburst' or 'bloom'.

The sudden decline of the phytoplankton crop, which marks the end of the spring bloom, has often been attributed to a decrease in the concentrations of inorganic nutrients, which occurs in many waters at around the same time of year (see Fig. 2.10). Although this explanation may be correct for some temperate lakes (e.g. the diatom crop in Windermere is limited by silicon[139]), it cannot always be applied to marine phytoplankton. Many results from temperate seas have shown that the crop declines before the decrease in inorganic nutrients occurs,[49] and this fall in inorganic nutrients may, in any case, be accompanied by a rise in organic sources of nitrogen and phosphorus, so that the change in the overall supply of these elements is small (see p. 75). However, the increase in the phytoplankton crop provides the food for, and is followed by, an increase in the zooplankton population, and it is this which — in the classic pattern of prey-predator relationships — brings the increase of the phytoplankton crop to an end, and starts the decline. The phytoplankton crop remains low throughout the summer in temperate waters (Fig. 4.8b), but the numbers of zooplankton do not show a corresponding decrease. This suggests that the phytoplankton are growing rapidly, and that their numbers are kept in check by zooplankton grazing, rather than by low nutrient availability. Since the water column is strongly stratified during this period, and inorganic nutrients are low (Fig. 2.10), phytoplankton growth must be supported by rapid recycling of nutrients within the photic zone (see pp. 69, 75). The recovery of the phytoplankton crop that often occurs for a brief period in the autumn is also correlated more closely with the decline in zooplankton numbers (Fig. 4.8b) than with the increase in inorganic nutrient levels that accompanies the decay of the thermocline (Fig. 2.10). The sudden return to fully mixed conditions in autumn will restore the nutrients to typical winter concentrations, but the depth of the mixed layer will also increase and will soon exceed the critical depth. Therefore, the collapse of the thermocline must mark the end, rather than the beginning, of the autumn peak in the phytoplankton crop.

In polar latitudes, winter temperatures will be low enough to inhibit the growth of both plants and animals in the plankton, even though the formation of ice will result in a more stable water column than is found in temperate seas during the winter. Snow covering the ice will prevent what little light there is from reaching the water, and so the spring bloom cannot occur until the ice breaks up in late spring. By this time, the daylength is longer than in temperate latitudes and the phytoplankton can grow very rapidly, but zooplankton numbers are so low at the end of the long winter season that there is a greater delay period between the increase of the phytoplankton and that of the zooplankton (Fig. 4.8a). This permits the algae to reach a higher peak of biomass than is often observed in temperate seas, but the timing of the peak, around midsummer, means that a secondary peak cannot occur before the ice re-forms in autumn.

In tropical seas, on the other hand, the amplitude of the seasonal variations in abundance of both phytoplankton and zooplankton is much less than at higher latitudes (Fig. 4.8c). Temperature and light conditions permit rapid phytoplankton growth throughout the year, and consequently there is always a substantial zooplankton population, which contributes to the recycling of essential nutrients. However, the phytoplankton crop is continually exposed to intense grazing pressure, and rarely produces the conspicuous blooms that are characteristic of higher latitudes.

Benthic algae

Species of *Laminaria* and other genera of kelps with single blades (e.g. *Saccorhiza, Alaria*) possess a unique advantage for studies of plant growth under natural conditions in that growth is confined to a single meristem located near the base of the blade. The oldest tissue, therefore, occurs near the tip of the blade, and the youngest near the base. Kelp plants are not systematically grazed by animals, although sea urchins frequently detach whole plants from their substrate by gnawing through the stipe (see p. 131), but the older tissue at the tip of the blade is continually worn away by water movement in the sea. Thus, the kelp blade can be thought of as a moving belt of tissue, whose length at any time is governed by the balance between growth at the base and erosion at the tip, and the rate of growth can be estimated simply by following the progress of markers (usually small holes punched through the blade) as they are pushed away from the basal meristem.

This technique has been used to study the growth of several species of *Laminaria* in temperate regions of the northern hemisphere, and has revealed a distinctive pattern of seasonal growth. Most species grow rapidly during the first six months of the year (January – June), but show a marked decrease in growth rate at around midsummer, and this slow growth continues for the rest of the year (Fig. 4.9b). The low summer growth rates were originally attributed to high respiration rates induced by high summer temperatures, but Lüning[140] has shown that the respiration of *L. hyperborea*

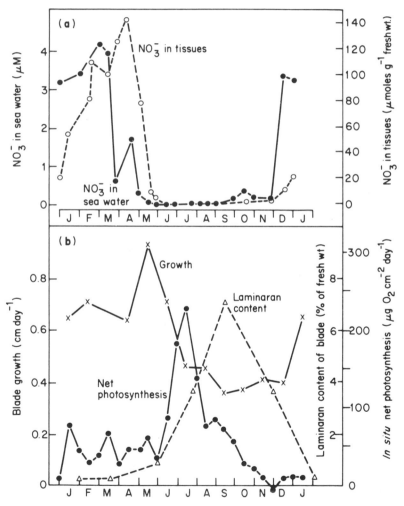

Fig. 4.9 Seasonal variations in the growth of *Laminaria longicruris* at a water depth of 9 m in St. Margaret's Bay, Nova Scotia, in relation to internal and external factors.[30, 31, 92] (a) NO_3-N in the sea water (●) and in the tissues of the plants (○); (b) growth in length (×), net photosynthesis *in situ* (●) and laminaran content of kelp blades (△).

does not increase with the seasonal temperature (see p. 79) and that net photosynthesis reaches a maximum in August, after the fall in growth rate. Another possibility is that kelp growth becomes limited by low external nutrient concentrations during the summer, but the sudden decline in growth rate does not occur until 2 – 3 months after the spring decrease of nitrate levels in the sea.[30] During the spring, however, kelp plants have been found to contain large reserves of nitrate in the tissues (up to 28 000 times the maximum sea water concentration), and these reserves remain high for about two months after the decrease in external nitrate (Fig. 4.9a). The fall in growth rate thus coincides with the reduction of internal nitrate to undetectable levels. The net photosynthesis of the plants remains high, however, and the excess photosynthate accumulates as reserve polysaccharide (laminaran, Fig. 4.9b). These carbohydrate reserves reach a peak in autumn, but decrease rapidly during the winter, since they are translocated to the meristem and support growth when light and temperature conditions are unfavourable, but nitrate levels are high again.[92, 140] Thus, the ability of kelp plants to store substantial supplies of nitrate when this nutrient is readily available in spring, and substantial reserves of carbohydrate when light and temperature conditions are favourable in late summer and autumn, alleviates the effects of nutrient-limitation or light-limitation at other times of the year, and probably accounts for the fact that kelp forests are among the most productive vegetation types in the world (see p. 89).

PRIMARY PRODUCTIVITY IN THE SEA

The main objective of studies of primary productivity is to measure the net production of all the plants in an ecosystem, since this determines the total amount of energy that is available to support heterotrophic growth. Estimates of autotroph respiration and gross primary productivity are of interest, if they can be obtained, but they are of secondary importance. Net primary productivity is equivalent to the rate of net photosynthesis per unit of biomass (averaged over 24 hours) multiplied by the standing crop, but it is not necessarily the same as the rate of growth of the crop, since some of the fixed carbon may be excreted as dissolved organic compounds. Plants of widely different morphology, growing in different marine habitats, all contribute to the primary productivity of the sea, and the relative importance of these different contributions, and the total productivity of the seas as a whole, can only be assessed by converting all productivity estimates to the same units. This process of conversion disguises the fact that different techniques must be used for different plants, and that the results obtained with different techniques may not be strictly comparable.

Techniques for estimating the productivity of marine plants

There are basically two distinct approaches to estimating primary productivity: measuring the exchange of individual chemicals between plants and their environment; and measuring the increase in biomass of a plant population. The first approach is usually based on carbon or oxygen measurements, and thus provides an estimate of *in situ* photosynthesis. It is, therefore, theoretically preferable to measurements of biomass, which, at best, estimate *in situ* growth. However, a major disadvantage of the chemical approach is that, in most marine situations, the plants (planktonic or benthic) must be enclosed for a period of hours in a transparent container, and this may result in significant changes in the environmental conditions, even if the plants are incubated at the site of their collection. Nutrient uptake is liable to be reduced, because the absence of water movement will permit a boundary layer of reduced concentration to build up around the cells, or over the surface of a thallus. The boundary layer around phytoplankton cells will normally be thin, even in completely calm conditions, because the cells move through the water as they sink, but this will be prevented in a bottle. The exclusion or reduction of the animal population, either deliberately to prevent grazing or as an accidental consequence of the sampling, may also reduce nutrient availability by preventing the rapid recycling of nutrients which is now thought to occur in natural waters (see p. 75). Another effect of enclosure on planktonic organisms is that cells will be exposed to constant irradiances for much longer periods than if they were circulating freely in the water. One possible result is that the inhibitory effects of high irradiances near the surface will be exaggerated, since inhibition increases with the length of exposure[87] (see p. 57). The walls of the enclosure also provide a new surface in the sea, which may encourage the development of unnaturally large populations of surface-living organisms, such as bacteria, whose respiration may significantly affect the productivity measurements. For all of these reasons, it is preferable to keep incubation periods as short as possible.

The **oxygen technique** for estimating productivity simply involves measuring the dissolved oxygen content of the water at the start and the end of the incubation. If all or most of the natural heterotrophs in the water sample can be excluded, the net change of oxygen concentration in the bottle represents net photosynthesis by the plant population (i.e. net primary production) but, if significant numbers of animals or bacteria are present, the result will estimate the net production of the community, rather than that of the plants alone. The simultaneous incubation of a darkened water sample provides an estimate of dark respiration rate, which can be added to the net change to estimate gross primary productivity (given the dubious assumption that respiration is unaffected by light), but the respiration of any heterotrophs present cannot be separated from that of the plant population. The applicability of the oxygen technique is limited by the sensitivity of the chemical methods for measuring dissolved oxygen, and long incubations of 12 or 24 hours are necessary for waters containing small phytoplankton populations. Nevertheless, it is the least problematic technique for measuring phytoplankton productivity in eutrophic waters, and it can also be applied to macroscopic marine plants (e.g. *Laminaria*,[92] *Macrocystis*,[246] seagrasses[57]).

The **radiocarbon technique** provides a more sensitive method for estimating phytoplankton productivity, but this advantage is counter-balanced by a number of disadvantages. The method involves incubating a sample of sea water, complete with its natural phytoplankton population, for $1-6$ hours with a small amount of radioactive bicarbonate ($H^{14}CO_3^-$) that is insufficient to produce a significant increase in the total CO_2-content of the water. After incubation, the phytoplankton cells are filtered off, and their radioactivity is measured to determine the proportion of the added ^{14}C that has been incorporated into the cells. The total production is then estimated by assuming that the cells have incorporated the same proportion of the total carbon in the water:

$$\text{i.e. } {}^{12}C \text{ assimilated} = \frac{{}^{14}C \text{ assimilated}}{{}^{14}C \text{ available}} \times {}^{12}C \text{ available}$$

(see Hall and Moll[86] for futher details). The sensitivity of the technique means that incubation times can be shorter than for the oxygen technique but, after 30 years of extensive use, it is still not clear whether the results represent gross photosynthesis, net photosynthesis, or some intermediate value. This is because it is possible that the carbon fixed in photosynthesis is released as respiratory CO_2 within the length of a single incubation, and that this CO_2 is immediately re-assimilated. However, the extent of such recy-cling of CO_2 can only be guessed.[87] Another problem is that carbon which is fixed and excreted as soluble organic compounds (see p. 77) will not be retained on the filter and counted as assimilated carbon, even though it con-tributes to energy flow in the sea via bacterial utilization (see p. 171). Further treatment of the filtrate permits radioactive organic carbon to be distin-guished from residual $H^{14}CO_3^-$,[86] but appropriate techniques have been stan-dardized only recently, and many estimates of primary productivity using the radiocarbon technique make no allowance for excreted photosynthate.

These differences between the oxygen and radiocarbon techniques mean that the results obtained by the two methods are difficult to compare directly, and any such comparisons are further complicated by uncertainty about the **photosynthetic quotient** (P.Q.). If the sole product of photo-synthesis is carbohydrate, the number of CO_2 molecules reduced in photo-synthesis will equal the number of oxygen molecules released but, since many compounds required for growth are more highly reduced than carbohydrates, the molecular ratio between O_2 and CO_2 (the P.Q.) will normally be greater than 1.0. Some results indicate that the P.Q. for phytoplankton is $1.2-1.3$, but the exact value will be affected by environmental conditions and by growth rate, and will probably also depend on the species composition of the phytoplankton crop. It should be clear, therefore, that neither of these superficially attractive techniques represents an ideal way of measuring primary productivity, and the alternative approach of measuring changes in biomass can yield results that are no less accurate (or no more inaccurate), but which are more economical in terms of research time and expense.

Changes in standing crop can provide a reliable estimate of population growth only if allowance can be made for losses due to grazing, sinking or water movement, or if such losses can be prevented. Unfortunately, enclosure is the only solution to this problem for phytoplankton, and the incubation times required to detect significant growth in natural populations (3 – 4 days) are unacceptably long. The growth of macrophyte populations is easier to study in the field, since grazing losses are usually small for established plants (although sporeling stages are heavily grazed; see Chapter 6), but whole plants or parts of plants will be continually removed by wave action. It is clearly impossible to prevent such erosion without elaborate and unnatural protection of the plants but, since the best growth season usually coincides with calmer conditions, reliable estimates of growth can be obtained by harvesting at carefully selected times of year (e.g. *Laminaria*,[118] fucoids[23]). For *Laminaria* species, it is relatively easy to measure the growth in length of individual fronds (see p. 83), but the conversion of these results to the growth in biomass of a whole population is far more difficult, and this approach offers little advantage over direct harvesting.[119] Annual production estimated from biomass changes or from *in situ* growth measurements has been directly compared with *in situ* photosynthesis throughout the year in a subtidal *Laminaria* population[92] and in intertidal fucoids.[23] Both of these studies showed that a large proportion (35 – 50%) of the annual net photosynthetic production could not be accounted for in the observed growth of the plants. This fraction presumably represents production that was lost as soluble organic compounds (possibly through the mucilages that are produced by many large brown seaweeds), and these results emphasize that all productivity estimates based on growth measurements or standing crop harvests will fail to measure this aspect of primary production.

The difficulties that have been experienced in measuring primary productivity in the sea have encouraged many attempts to formulate mathematical models of the process. The aim of these **productivity models** is to predict photosynthesis or productivity at frequent intervals from continuous and automated measurements of chlorophyll concentration (as an estimate of standing crop) and major environmental factors, such as surface irradiance, transmittance of light through the water column, temperature and nutrient concentrations. To date, the modelling approach has been no more successful than other approaches to productivity, partly because reliable models must be built upon a firm theoretical base. This demands a better understanding than is currently available about the interactions between different environmental factors in their effects on growth and photosynthesis, and about the adaptation of plants to the prevailing environment. Model development has also been inhibited, however, by the lack of any effective means of validating existing models. In the absence of a really satisfactory field technique for estimating productivity, any discrepancy between model predictions and field estimates could be attributed to errors in the model, or to the failings of the field technique — or, of course, to both. Even worse for the modeller's morale is the thought that agreement could simply mean that both model predictions and field estimates were in error by the same amount! The available estimates of marine productivity are best viewed through the sceptical haze produced by these somewhat disturbing thoughts.

Estimates of primary productivity in marine ecosystems

A reasonably complete picture of the productivity of marine phytoplankton throughout the world has been built up over the last 30 years (Fig. 4.10), which is largely based on results from the radiocarbon technique. Although these results may need to be modified in the light of criticisms of particular aspects of the technique (see p. 87), it is unlikely that the overall pattern of low productivity in tropical oceans and higher productivity in shallower, temperate waters and in areas of upwelling will be substantially changed. There is some evidence, for example, that oceanic productivity has been underestimated by the standard techniques,[71] and that phytoplankton growth is not so strongly limited by nutrient supply as has long been thought (see p. 75), but the small amount of chlorophyll per unit surface area in most oceans (10% of that in coastal waters, and only 1% of typical forests[259]) must mean that oceanic productivity will be lower than in most other ecosystems. Benthic macrophytes growing around the edges of the sea intercept more of the incident light because of higher chlorophyll concentrations, and because, by living at a fixed depth, they are exposed to a more constant environment than phytoplankton. Storage of nutrients or carbohydrate by benthic plants also means that the effects of seasonal changes in the environment are partially alleviated (see p. 85), and consequently such plant communities achieve rates of production that are higher than any phytoplankton populations, and are similar to some of the most productive terrestrial vegetation types (Fig. 4.11).

Estimates of the total area of sea surface occupied by each type of marine vegetation can be combined with productivity data to calculate the total primary production in the sea. On a surface area basis, about 90% of the world's oceans contain the least productive phytoplankton populations, and the highly productive benthic plants and phytoplankton of upwelling areas occupy such a small proportion that they contribute relatively little to the total marine production (Table 4.3). This has important implications for the animal life that the sea can support, and for the total yield of fish that man can expect to obtain from the sea. Ryther[219] has calculated, on the basis of the primary productivity estimates in Table 4.3, that the total annual production of fish in the sea is about 24×10^7 tons. The estimated fish production in oceanic waters represented less than 1% of the world total, because oceanic phytoplankton consists mainly of small-celled species, and longer food chains must intervene between phytoplankton and catchable fish in the open ocean than in coastal and upwelling areas.[219] Substantial upward revision of the current estimates of primary productivity by oceanic phytoplankton (even by as much as 10 times) will, therefore, have little influence on the estimates of total fish production, and none of the other primary productivity estimates in Table 4.3 is thought to require significant

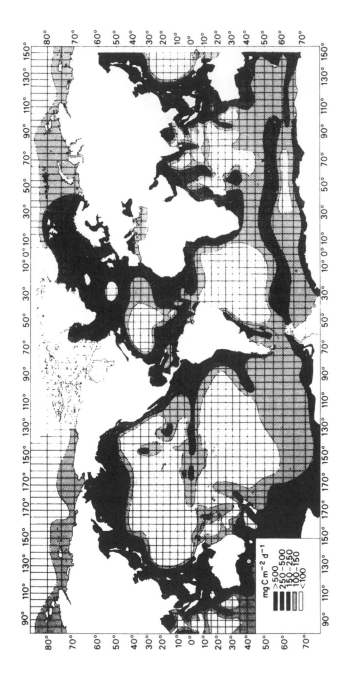

Fig. 4.10 Primary productivity of phytoplankton in the world's oceans. (From Map no. 1.1, *Atlas of the Living Resources of the Sea.* Reproduced by permission of the Food and Agriculture Organization of the United Nations, Rome, 1972.)

mg C m^{-2} d^{-1}
>500
250–500
150–250
100–150
<100

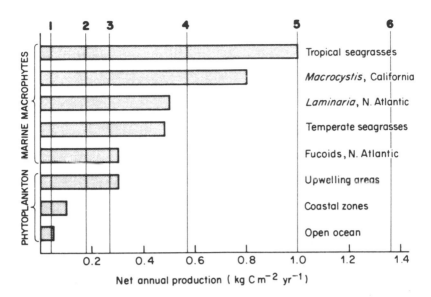

Fig. 4.11 Net annual production of marine vegetation types, compared with terrestrial vegetation (1 = desert scrub; 2 = lake and stream; 3 = temperate grassland; 4 = temperate forest; 5 = tropical rain forest; 6 = swamp and marsh). (Redrawn from Mann,[153] © 1973 by the American Association for the Advancement of Science; with data concerning terrestrial vegetation,[259] seagrasses,[158] Macrocystis,[39] Laminaria,[119] fucoids[23] and phytoplankton.[219])

Table 4.3 Total net primary production in the world's oceans.[219, 259]

Region	Net primary productivity (g C m^{-2} yr^{-1})	Surface area (10^6 km^2)	Total net production (10^9 t C yr^{-1})
Open ocean	50	326	16.3
Coastal zone	100	36	3.6
Upwelling areas	300	0.36	0.1
Benthic ecosystems	600	2	1.2
Total marine	58	364	21.2
Total continental	358	149	53.4
World total	145	513	74.6

alteration. The total world harvest of marine organisms reached about 6 × 10^7 tons in 1970 and has increased very little since (see p. 156), so that the world's fishing industry appears to be close to the absolute limit imposed by the productivity of marine plants. The idea that the sea contains large, untapped reserves of food for human consumption must be dismissed.

5

Morphogenesis of Marine Plants

The discussion of morphology in Chapter 1 raised several fundamental questions about marine plants, such as why the algae should have developed more complex morphologies in the sea than in fresh water, and why, among photosynthetic plants, both pseudo-parenchymatous and siphonaceous thalli should be almost entirely confined to the sea. Another question that could be asked is why the marine algae as a group should show such an extraordinary diversity of life histories. The suggestion that the different morphologies, life histories and physiologies of marine algae represent 'evolutionary experiments' may, in a sense, be true, but it provides no more than a superficial answer to these questions. It fails to explain why these features have persisted — or continued to evolve — after the successful invasion of the land by parenchymatous plants, or why one of these marine experiments has not eliminated all others in the sea.

Current knowledge of the factors controlling algal morphology is far too rudimentary for us to be able even to see where the answers to such questions may lie, but the solution of these problems provides a long-term objective for studies of algal morphogenesis. Meanwhile, less far-reaching questions about how a particular environmental factor affects the morphology or the development of a particular species, or about the relative importance of genetics and the environment, provide the main stimulus for research. It will be clear from the work described in this chapter that studies of algal morphogenesis have been spread very thinly among the questions that could be asked, although some marine algae have been intensively investigated because, as with photosynthesis (see p. 43), they provide convenient experimental systems for tackling problems of wider biological significance. Examples include studies of nuclear–cytoplasmic interactions (*Acetabularia*[200]), the establishment of cell polarity (*Fucus* zygotes[202]), and the morphogenesis of cell walls (*Cladophora, Valonia* and other large-celled green algae[148]). Since such work does not relate specifically to marine problems, and is well covered by the references cited above, it is not discussed in detail here.

ALGAL LIFE HISTORIES: AN OUTLINE OF THEIR COMPLEXITY AND VARIETY

The life histories of all flowering plants, together with all gymnosperms, pteridophytes and bryophytes, follow a single basic pattern. There is always an alternation between a haploid, gamete-producing phase (the *gametophyte*) and a diploid, spore-producing phase (the *sporophyte*), which are both morphologically and cytologically distinct (Fig. 5.1c). In most of these plants, one of the phases is non-photosynthetic and is unable to grow independently of the other, but the ferns provide a familiar example of plants in which the two phases are so distinct from, and independent of, each other that the connection between them would not be suspected if the fate of the reproductive bodies that they each produce was not known. For precisely this reason, algal taxonomists have often been forced to draw red lines through familiar and well-established names in their floras, as one culture study after another has shown that two algae, long known as different species and usually classified in different families or even in different orders, were in fact two phases in the life history of a single species. Such discoveries started as early as the 1870s, when both the filamentous

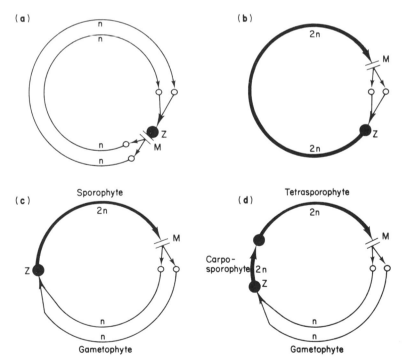

Fig. 5.1 Life history patterns of marine algae. Thin lines and open circles represent haploid phases; bold lines and filled circles represent diploid phases. M = meiosis; Z = zygote.

green alga *Ulothrix zonata* and the brown alga *Cutleria* were found to share their life histories with other 'genera' — *Codiolum* and *Aglaozonia*, respectively — but they are evidently far from complete. There has been a recent spate of similar reports, mostly concerning red algae.[19] It is now clear that such **heteromorphic** life histories are common in all three groups of macroscopic marine algae (see Table 5.1). Not all of the alternate phases, however, are conspicuous enough to have been described as separate species. The gametophytes of *Laminaria*, for example, are so small as to be almost undetectable in natural conditions, but the life history of these plants was shown, through culture studies, to be similar to that of the ferns as long ago as 1916.

Two variants on the life history pattern of the ferns can be found among marine algae, although not in all groups. Firstly, the two phases in the life history may have identical morphologies. In these **isomorphic** life histories, the two phases are cytologically distinct — one phase is haploid, and the other diploid — and they produce different types of reproductive bodies, but are otherwise indistinguishable from one another with, as far as is known, identical physiological and ecological characteristics, as well as the same morphology. This pattern is common in both green and brown

Table 5.1   Life histories of marine algae.

(a) Single morphological phase: somatic cells haploid or diploid

Group	Haploid	Diploid
Chlorophyta	*Dunaliella*	*Codium, Caulerpa, Udotea, Halimeda*
Phaeophyta	—	*Fucus, Sargassum* and other Fucales
Rhodophyta	—	—
Others	Dinoflagellates	Diatoms, some dinoflagellates

(b) Two morphological phases: haploid gametophyte and diploid sporophyte

Group	Isomorphic	Heteromorphic Gametophyte	Heteromorphic Sporophyte
Chlorophyta	*Ulva, Cladophora* *Enteromorpha*	*Monostroma, Acrosiphonia* *Halicystis, Bryopsis*	Codiolum *Derbesia*
Phaeophyta	*Ectocarpus,* *Dictyota*	*Scytosiphon, Petalonia* Laminarian gametophytes	*Ralfsia*-like *Laminaria*
Rhodophyta	—	*Porphyra, Bangia*	Conchocelis

(c) Three morphological phases (Rhodophyta only): haploid gametophyte, diploid carposporophyte developing on female gametophyte, and independent diploid tetrasporophyte

Isomorphic	Heteromorphic Gametophyte	Heteromorphic Tetrasporophyte
Chondrus, Delesseria	*Bonnemaisonia*	Trailliella
Eucheuma, Gracilaria	*Gigartina*	Petrocelis
Griffithsia, Polysiphonia	*Acrosymphyton*	Hymenoclonium

algae, but it has not so far been reported for a red alga, except when it is combined with the second variant, which is found only in red algae. The typical life history of the more complex and highly evolved red algae in the subclass Floridiophyceae contains two morphologically-distinct diploid phases, in addition to the haploid phase (Fig. 5.1d). The two diploid phases are known as the **carposporophyte** and the **tetrasporophyte**, but the former is never an independent plant and always develops on the female gametophyte. The two independent phases in the life history — the game-tophyte and the tetrasporophyte — may be indistinguishable, or they may be totally different in morphology, and it is here that isomorphic and heteromorphic types of life history can be identified among these algae (see Table 5.1).

In addition to these rather complex life histories with two or three morphological phases, marine algae also exhibit simple, single-phase life histories, more closely resembling those of animals than those of any other photosynthetic plants. Thus, the familiar species of *Fucus* and other related brown seaweeds are diploid, like the sporophytes of the Laminariales, but they produce gametes directly, rather than going through a gametophyte phase (Fig. 5.1b). Many of the larger siphonaceous green algae are also of this type, and the sexual cycle of diatoms follows a similar pattern. The simplest — and possibly most primitive — life history of all, in which the somatic cells are haploid and meiosis occurs when the diploid zygote germinates (Fig. 5.1a), is found in the marine flora only among some of the unicellular algae (Table 5.1) and a few species of multicellular green algae (e.g. *Rosenvingiella constricta*[126]).

The different types of life history exhibited by marine algae are sum-marized in Fig. 5.1 and Table 5.1, and the examples in the table include most of the plants which figure prominently in this book. Further details about all of these life histories are given by Bold and Wynne.[19]

GERMINATION AND ATTACHMENT OF SPORES

The first and possibly most critical stage in the development of every independent phase in the life history of an attached benthic alga is the germination of the spore or zygote, and the attachment of the young plant to a substrate. In spite of the extensive research devoted to seed and spore germination in almost all other plants, the germination of algal spores has received very little attention. Even in groups which have been the favourite subjects of experimentalists, such as the Laminariales and the Fucales, there have been few systematic investigations of germination under controlled conditions. The results that are available suggest, however, that algal spores will germinate rapidly under a wide range of conditions. Temperature, for example, has little effect on the percentage germination of either laminarian zoospores or fucoid zygotes from $2-3°C$ up to the lethal limit of these plants at $20-25°C$ (Table 5.2). They also appear to germinate equally well in either light or darkness (e.g. *Saccorhiza*,[180] *Laminaria*,[142] *Halidrys*[165]). The apparent absence of the rather specific germination requirements that are

Table 5.2 Percentage germination of zoospores of *Saccorhiza* spp. after 14 days at different temperatures.[180]

Species	Temperature (°C)						
	3	5	10	17	20	23	25
Saccorhiza polyschides	85	83	87	74	96	80	16
Saccorhiza dermatodea	74	88	86	98	76	60	2

often found in seeds is perhaps to be expected, since algal spores are generally smaller and presumably contain less food reserves which could be utilized during a period of suspended animation. Very few of the 'spores' produced by marine algae have the resistant walls and low metabolic rates found in most of the spores and seeds of terrestrial plants, and germination either need not, or cannot, be delayed until conditions are favourable.

Although the environment has little influence on the rate of germination, both the pattern and direction of germination may be strongly modified by the environment. The polarity of the developing embryos of *Fucus* and *Pelvetia* (i.e. which side becomes the rhizoid, and which the vegetative thallus) can be determined by any one of about fifteen physical, chemical or biological factors.[103] The most important factor for a zygote on the shore, however, is probably light, since the rhizoid develops on the shaded side of the cell, and will respond to a small photon dose[14] as low as 200 μmol m^{-2}. If the zygotes are irradiated from one side with polarized light, germination occurs from both the light and the dark poles of the cell (Fig. 5.2). This suggests that the pigment molecules which detect the light are located in the peripheral cytoplasm, and are oriented parallel to the cell surface. Such an arrangement would mean that the molecules at the sides of the cell absorb more light than those at either the illuminated or the shaded poles, so that both of these poles appear to be 'dark'. Germination, therefore, occurs from them both.

In the zygotes of all fucoid algae, the first visible sign of germination is the production of a rhizoid, and this emphasizes that attachment must occur early in the life of any spore. Rapid attachment is particularly important for species in the intertidal or upper subtidal zones that are subject to substantial wave action. Primary attachment generally occurs before the rhizoids

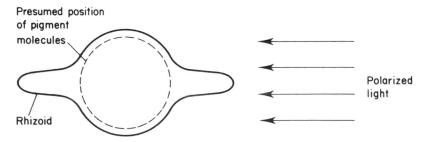

Fig. 5.2 Germination of a zygote of *Fucus* in polarized light.[103]

are produced, however, and the rate of attachment of algal spores has been estimated by simply counting the number that could be washed off a surface at different times after settlement. The apparatus used has often been very simple, and the precision with which the applied forces could be controlled or measured mostly very low (e.g. a mains water tap, with or without some control of water pressure), but Charters et al.[35] developed a 'water broom', which simulated remarkably well the complex of forces that a wave exerts on a rock surface (Fig. 5.3). Unfortunately, few detailed results were obtained with this apparatus, but it was shown that, 7 hours after settlement, 50% of the tetraspores of the red alga *Gracilariopsis* could withstand dislodging forces that were 100 times the weight of the individual spores. Seven hours is a long time on a wave-battered shore, but most of the less elaborate work on attachment has confirmed that it generally takes at least as long as this for a substantial percentage of algal spores to become firmly attached. What has not been established, however, is whether a spore is permanently lost once it is detached, or whether it continues to develop its 'stickiness', so that it can attach to another substrate if and when the opportunity arises.

The intensive work on polarity in fucoid zygotes has thrown some light on the biochemistry of attachment. It was soon noticed that fertilized and unfertilized eggs could be separated before any visible sign of germination, because the unfertilized eggs did not become attached to the substrate. Thus, attachment clearly precedes rhizoid emergence and, by following germination in a suspension of non-toxic particles (e.g. resin beads) in sea

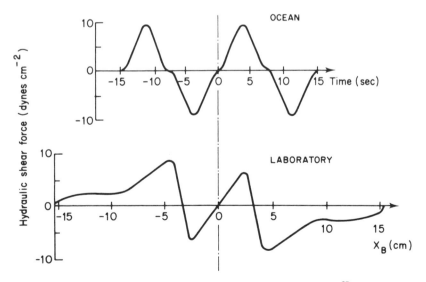

Fig. 5.3 Comparison of surface shear in the ocean and the laboratory.[35] (*Upper graph*) Force at the centre of a rock 2 m long, produced by a wave with a period of 15 s and maximum velocity of $0.6 \, m \, s^{-1}$. (*Lower graph*) Force generated on a substrate surface by a 'water broom', designed to measure spore adhesion in the laboratory.

water, it can be shown that the zygote secretes a greater thickness of jelly at the point where the rhizoid is due to emerge some three hours later.[226] If the zygotes are germinated in the absence of sulphate, however, no jelly develops and no attachment occurs, even though rhizoid development is perfectly normal. This result appears to be related to the observation that fucoidan — a sulphated polymer of the simple hexose sugar fucose — is localized in the rhizoid cell of normal embryos, but it is not formed in the absence of sulphate.[47] The polysaccharide fucoidan, therefore, appears to be responsible for sticking fucoid embryos to their substrate. The attachment mechanisms of the green alga *Enteromorpha* and the brown alga *Ectocarpus* have also been studied intensively because of the importance of these species as ship-fouling algae. *Enteromorpha* zoospores were subjected, during or after settlement, to various enzymes, such as amylase, trypsin and pronase, and the degree of detachment by a water jet was measured. All three enzymes appeared to weaken the adhesion of the zoospores, and subsequent electron microscope autoradiography confirmed that the adhesive contained both protein and polysaccharide. This was synthesized in the Golgi apparatus and secreted to the outside of the cell by exocytosis within minutes of initial contact between the zoospore and a substrate.[37]

These adhesives, whatever their chemical nature in different algae, provide a temporary attachment for the sporeling before the rhizoids appear (a period of several days in some fucoids), and they continue to be secreted by the rhizoids as these develop. The strength of the sporeling's permanent attachment will clearly be related to the total length of its rhizoids. Some fucoids produce large numbers of primary rhizoids (e.g. *Himanthalia*, Fig. 5.4), while others show rapid rhizoid elongation, even in complete darkness (e.g. *Halidrys*[165]). Prolonged darkness (> 40 days) causes growth to cease, but *Halidrys* sporelings were able to resume growth when returned to the light after as long as 120 days in the dark. This ability of sporelings to attach themselves and then survive a prolonged period of darkness must be of considerable ecological significance in a subtidal species that reproduces in winter, such as *Halidrys*. Gametophytes of all three European species of *Laminaria* can also survive in complete darkness for at least 5 months.[142] The zoospores germinate to produce the primary cell of the gametophyte, but do not develop any further until they receive sufficient light for photosynthesis. These observations disprove the hypothesis (p. 96) that algal spores have insufficient reserves for a prolonged period of inactivity. In the marine environment, however, it is essential that attachment should precede any such period of enforced dormancy. The rapid germination of algal spores under most environmental conditions may, therefore, be interpreted simply as a method to ensure attachment. Zygotes of *Fucus* can germinate (i.e. produce a rhizoid) without attaching (see above), but there is no evidence that spores can attach without germinating soon afterwards.

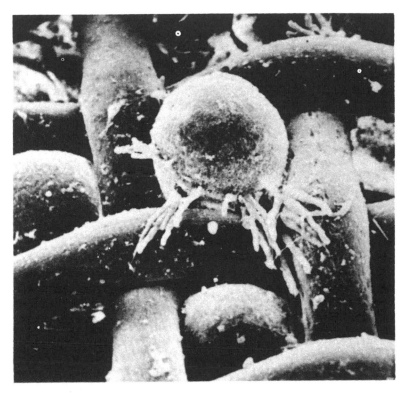

Fig. 5.4 Zygote of *Himanthalia elongata* germinating on nylon fabric. Numerous primary rhizoids have been produced where the zygote is in contact with its substrate. (Scanning electron micrograph × 200, by courtesy of Dr. B.L. Moss.)

ENVIRONMENT AND VEGETATIVE MORPHOLOGY: OBSERVATIONS AND EXPERIMENTS IN THE FIELD

Large seaweeds are difficult and inconvenient organisms to grow to maturity in culture, and the impossibility of reproducing many aspects of their natural environment (e.g. wave action) in a culture chamber must mean that the conditions are highly artificial, and the results are correspondingly suspect. The best approach to the study of seaweed morphogenesis, therefore, may be to let nature do the experiments, and to look for correlations between morphology and environment in the field. The appearance of a species in two different habitats is sometimes so distinct that individual plants from different sites have been described as two separate species, which have been united only after the intermediate forms have been found in intermediate habitats, and plants moved from one habitat to the other have been shown to change their morphology in response to the change in environment.

The factor that has received most attention, and which produces the most striking effects in this connection, is exposure to wave action. Most of the species of kelps in the North Atlantic show remarkably similar morphological responses to an increase in exposure: the blade becomes narrower and thicker, the stipe becomes shorter and stronger, and the broad blades of species such as *Laminaria digitata*, *L. hyperborea* and *Saccorhiza polyschides* become split into more and more 'digits'. Some of these responses are effectively demonstrated by a single plant of *Alaria* (Fig. 5.5) that was transplanted from an exposed to a sheltered location. The new tissue produced at the base of the blade (see p. 83) reflects the characteristics of its new habitat, whereas the older tissue towards the tip was produced in the original habitat. Some of the fucoids (e.g. *Fucus vesiculosus*, *Pelvetia canaliculata*) also show markedly different morphologies in sites with different exposure. However, the morphological pattern of the fucoids, with apical growth and both dichotomous and lateral branching, is so much more complex than that of the laminarians that the effects of environment on morphology are more difficult to quantify and to separate from concurrent changes in growth rate. For example, the number of dichotomies per unit fresh weight of thallus in *Fucus vesiculosus* increased with exposure (Fig. 5.6) but, since the size of the plants decreased along the same environmental gradient, these results may mean that the number of dichotomies *per plant* was little affected by exposure. The number of air-bladders in the thallus of this species also seems to decrease with increase in exposure but, again, it is difficult to know how best to express this observation quantitatively (e.g. number of bladders per plant, or per unit length of thallus, or per unit area, or weight). In order to examine the relationship between morphology and exposure in detail, exact measurements of morphology are needed at a series of sites along a measured gradient of exposure (see p. 39). Unfortunately, most investigators have been satisfied with a comparison of plants from two extreme sites, and Fig. 5.6 represents a rare attempt to improve on the simple observation that plants growing in exposed sites differ from those in sheltered sites.

Other environmental factors which have been shown to influence the morphology of marine algae under natural conditions include height on the shore, salinity, and whether or not the plant is attached. Towards the upper limit for *Ascophyllum* in the intertidal zone, individual plants are shorter than those growing lower on the shore, and have smaller air-bladders and

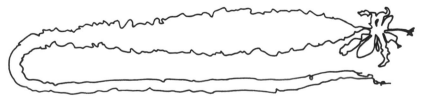

Fig. 5.5   Plant of *Alaria esculenta* (× 0.15) six months after transfer from an exposed site (west coast of Norway) to a sheltered site (Oslofjord). Drawn from a photograph by Sundene.[242]

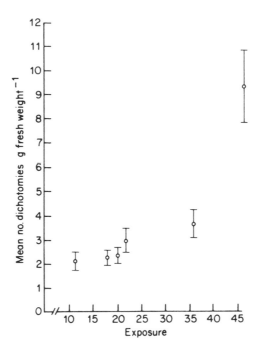

Fig. 5.6   Degree of branching in random samples of *Fucus vesiculosus* from several sites of different exposure on the Isle of Man.[217] Exposure values were calculated using Baardseth's method (see p. 40), weighted according to the prevailing wind patterns.

broader, thinner internodes. Another fucoid, *Halidrys*, exhibits a different pattern of branching when growing in intertidal rock-pools, compared with plants from the subtidal zone.[217] However, a large number of environmental factors are correlated with height on the shore (see pp. 34–9), and it is impossible to establish, by field work alone, the exact causes of such changes in morphology. Salinity represents a single environmental factor which has been shown to affect several aspects of the morphology of *Fucus vesiculosus*. The number of air-bladders per plant and the degree of branching increase in regions of low salinity,[113] and so does the weight of reproductive tissue, as a proportion of the total weight of the plant.[218] However, as with so many studies of exposure, only two extreme sites were investigated, and the effects of intermediate salinities were not elucidated.

A final example of the influence of environment on algal morphogenesis concerns two situations with no intermediates. Russell[213] noticed that one rather rare British species of *Ectocarpus* (*E. distortus*) had always been reported from drift material, tangled up with other plants or cast up on the beach, whereas the much commoner species *E. fasciculatus* had never been found in the unattached (or 'free-living') state. He began to suspect that these plants might be the same species, in spite of the marked differences in morphology. Plants of *E. fasciculatus* were, therefore, detached from their natural substrates and grown in the sea in transparent chambers attached to the mooring chain of a navigation buoy. The continuous movement meant that the plants were never able to re-attach themselves and, within 7 weeks, they had developed the typical morphology of *E. distortus*. The detached plants lost the polarity that was apparent in attached individuals, and this contributed to a change in the pattern of branching and in the length of branches that were formed. There were also changes in the position and size of the reproductive structures. Substantial alterations in many environmental factors are clearly involved in the change from the attached to the free-living state, and exactly what caused each of the morphological changes in the plants is not known. Nevertheless, experiments such as this provide a valuable source of ideas for detailed morphogenetical investigation in the laboratory.

MORPHOGENESIS OF ALGAL THALLI: LABORATORY STUDIES

Marine algae offer unique opportunities for the study of plant morphogenesis because of particular types of thallus construction that occur only in these plants. The large cells of certain green algae, such as *Acetabularia*, *Valonia* and *Cladophora* have been well exploited in such work for many years (see p. 92), but the pseudo-parenchymatous thalli of the more complex red algae (Florideophyceae) have received far less attention. The morphogenetical interest of these plants lies in the fact that the thallus is constructed almost entirely of uniaxial filaments, in which division is confined to the apical cell. The cells behind the apex may divide laterally to initiate a new secondary filament (in which, again, all growth will be apical), but they never divide transversely to add cells to the primary filament. This has a number of important implications:

(1)   the *number* of cells along the main axis of a filament is a measure of the total number of divisions that the apical cell has undergone;

(2)   the *size* of any cell in a filament, less that of the sub-apical cell (i.e. the cell that has just been formed), is a measure of the total enlargement of the cell since its formation;

(3)   the *position* of a cell in a filament is a measure of its relative age, and this can be converted to an absolute age if the rate of division of the apical cell is known (from (1));

(4)   the *sequence* of cell sizes along the filament represents the pattern of cell enlargement within the filament.

If single cells, about 1 mm in length, are isolated from shoots of the large-celled red alga *Griffithsia pacifica*, they rapidly regenerate into a small plant with the normal morphology of the species. Under favourable conditions, the apical cell of the main filament divides 10 – 12 times in 8 days, and the apical cell of each branch also divides at a similar rate once each is formed (Fig. 5.7). Each cell elongates from about 0.12 to 1.0 mm in 10 days, and then stops growing.[66] This represents a 30 – 40-fold increase in volume, but the cells of other red algae are very much smaller initially, and their final size

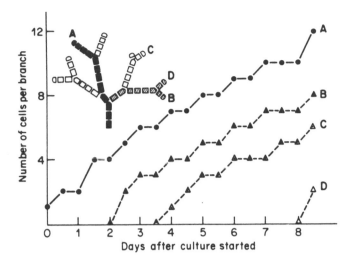

Fig. 5.7 Pattern of cell division in individual branches of a plant of *Griffithsia*, regenerating from a single isolated shoot cell.[66]

may be as much as 48 000 times their initial volume (e.g. *Antithamnion plumula*[54]). In angiosperms, an increase in cell volume of 200-fold is regarded as exceptionally large. Analysis of the pattern of cell enlargement shows that cells rarely enlarge at all points along their length. In most red algae, elongation occurs mainly towards the base of the cell, and — as a result — branches appear to arise from the apical end of axial cells. The branches are actually formed near the middle of young axial cells but, since subsequent elongation is confined to the basal half of the parent cell, the branch appears to become more and more apical in position. These patterns of cell enlargement can be elegantly revealed using a fluorescent dye that stains cell walls.[255] Living plants are stained for 15 – 30 minutes, and then allowed to grow on in the absence of the dye. The new cell wall material shows up as dark bands in the older stained walls (Fig. 5.8), and this confirms that wall synthesis is localized at the base of the cells in most red algae. In *Griffithsia*, however, bands of new wall are formed at both ends of the cells; this pattern of 'bipolar band growth' seems to be unknown in any other plants.

By use of data collected in this way, the total growth of red algae can be separated into two components — cell division and cell enlargement — and the responses of each component to different environmental factors can be studied. The growth of *Griffithsia* and *Pleonosporium* in different irradiances and daylengths[172] indicated that both apical cell division and cell enlargement in the main axis were light-saturated at low total daily irradiances (620 mmol m⁻²) but, in *Griffithsia*, the production of branches was strongly inhibited at low irradiances (Table 5.3) or in short daylengths.[252] Thus, the main axes of plants grown in low light were similar

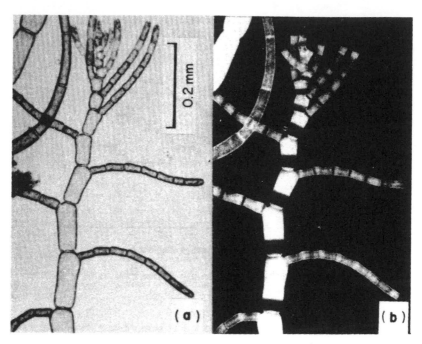

Fig. 5.8 Cell elongation in *Callithamnion*: the plant was stained for 30 min, and photographed 24 hours later (a) under normal illumination, and (b) using ultra-violet light, which causes the stain to fluoresce.[255] Cell wall material formed since staining shows up as dark bands in (b).

Table 5.3 Effect of irradiance on apical cell division and branch formation in single-cell regenerates of *Griffithsia pacifica*. Plants were grown for 7 days; each value is the mean of 5 plants.[252]

Photon irradiance (μmol m^{-2} s^{-1})	Number of branches per plant	Number of cells per primary filament
8	0.8	7.0
60	8.2	8.2

in length and cell number to those of plants grown in high light, but the latter were well-branched and bushy, whereas the former had no branches at all. Such changes in the balance between the growth rate of the main filament and that of the branches have major effects on the overall morphology and appearance of adult plants with pseudo-parenchymatous thalli, since the very solidity of the plant depends on the development of lateral branches, which surround and strengthen the main axis. In *Ceramium*, for example, the large cells of the central axis are covered by a layer of small cells (*'cortication'*), which is produced by the outgrowth and fusion of secondary filaments from the axial cells. When one species was grown in

8-hour days, the growth of the corticating filaments matched the elongation of the axial cells, and a continuous layer of cortication was formed. In 16-hour days, however, axial cell elongation was more rapid and the cortication was reduced to bands covering only the junctions between the axial cells.[82] Since the distinction between continuous and banded cortication has been regarded as an important taxonomic character in this genus, it looks as though these results will add to the problems that modern culture work is causing for algal taxonomists (see also pp. 93, 101).

Growth substances in marine algae

By measuring all the axial cells in the main filament and the major branches of mature thalli of red algae, Dixon[54] obtained evidence that the final size of cells was controlled by internal factors, as well as by the external environment. Cells at the base of lateral branches are sometimes only half the length of younger cells formed 5 – 10 divisions later, but located further away from the influence of the main axis (Fig. 5.9b), and the cells of the main axis are also smaller at and just above the points where the major lateral branches arise (Fig. 5.9a). These signs of interactions between cells within a single thallus, combined with observations of apical dominance in the thalli of both red and brown algae, led to the suggestion that substances similar to higher plant hormones — 'growth substances' — might be involved in the control and integration of growth in marine algae. Although this suggestion has been generally accepted in principle, there has been much dispute about the identity of these algal growth substances, and about the extent of their similarity to those in flowering plants. There have been numerous reports of auxins, cytokinins and gibberellins being isolated from marine algae, but most of these require re-investigation because neither the chemical analyses nor the bioassay techniques used were sufficiently specific. The equally numerous reports of the growth or morphology of marine algae being affected by exogenously-applied hormones have also been criticized because

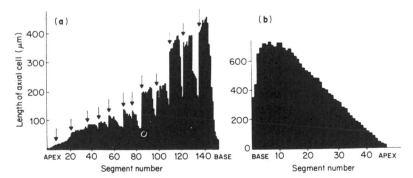

Fig. 5.9 Lengths of every axial cell in filaments of two red algae.[54] **(a)** Principal axis of *Ceramium* — arrows indicate positions of lateral branches; **(b)** primary lateral axis of *Callithamnion*.

they fail to demonstrate that these substances are essential for normal growth and development.

The best evidence that some macroscopic algae require specific organic 'growth substances' for their normal morphogenesis is provided by the observation that certain species develop totally abnormal morphologies when grown in a completely-defined inorganic medium in the absence of bacteria. The original experiments of this type involved the green alga *Ulva*, which developed what was picturesquely described as a 'pin-cushion morphology' instead of the familiar leafy thalli. Although certain specific combinations of growth substances did induce some of the plants to become a little more normal, the most striking morphogenetic effect was produced by adding strains of bacteria that had been isolated from *Ulva* plants in the field. Sterilized extracts of such bacteria were sometimes effective in restoring the morphology of similar cultures of *Monostroma* and *Enteromorpha*, but *Ulva* appeared to require growing bacteria in the culture for any development to occur.[199] It is still not clear whether the bacteria themselves produce the necessary morphogenetic substances, or whether they stimulate the algae to produce them. The identity of these substances is also in doubt, but it seems certain that compounds similar to, but not necessarily identical with, the hormones of vascular plants do play a role in the morphogenesis of marine algae. Meanwhile, a new ecological dimension has been added by the report that cytokinin-like activity can be detected in sea water collected from the *Fucus-Ascophyllum* zone of the intertidal, since such water has long been known to support better algal growth in cultures than water collected from further off-shore.[188]

The role of light in algal morphogenesis

Further parallels between marine algae and other plants can be seen in their morphogenetic responses to light. The attaching organs of several species of algae grow away from a unilateral light source, and so display the same negative phototropism as the roots of vascular plants. The response can be seen most clearly in the holdfasts of many kelp species, since the individual haptera, which make up the holdfast, can be induced to grow upwards if illuminated from below (Fig. 5.10). Phototropism in algae has not been studied as intensively as in vascular plants and fungi, but the haptera of *Alaria* and the rhizoids of *Griffithsia* are both similar to organs showing phototropism in other plants in that they respond to blue light but not to red.[25, 254]

A different type of response to blue light, which also seems to be related to attachment, invites comparison with another well-studied morphogenetic system in vascular plants. The brown alga *Scytosiphon* is commonly found as tubular erect thalli in the lower intertidal zone of many temperate shores around the world, but it is represented in the sporophyte generation by small crusts on the surface of rocks and boulders (previously described as *Ralfsia*; see Table 5.1). However, zoospores released from the erect plants develop into typical *Ralfsia*-like crusts only if they receive blue light. In red light,

Fig. 5.10 Negative phototropism in holdfasts of *Alaria*.[25] The plants were illuminated through the bottom of the culture tank, and all the haptera grew away from the light source, regardless of the orientation of the stipes.

they form irregularly-branching filaments which do not attach the plant to its substrate (Fig. 5.11A). This response to blue light can be detected in sporelings that are only 4 – 5 cells long, since those in blue light are broader than those in red, and the apical cell first becomes heart-shaped (Fig. 5.11B), and then divides longitudinally to initiate the 2-dimensional growth that will result in the crust.[62] This response is remarkably similar to the photomorphogenesis of fern gametophytes, which also remain filamentous in red light[161] but, because light quality under water is more variable than on land (see pp. 14 – 17), the algal response may have greater ecological significance than that in the ferns. If zoospores of *Scytosiphon* settle in relatively deep coastal water, they may receive too few blue quanta (see Fig. 2.3) to induce the formation of crusts, so that the plants can become established only in the intertidal or upper subtidal zones.[61] This response illustrates how the vertical distribution of a species can be controlled by the spectral distribution of underwater light, and also how morphogenetical investigations of marine plants in the laboratory can lead back to, and tie in with, observations in the field, with which they may have begun (see p. 102).

REPRODUCTION

Two types of reproduction are commonly described in plants — sexual and asexual — but the complex life histories of many marine algae include several types of asexual reproduction, and some expansion of this category is required. In the following discussion, most of the anatomical and structural

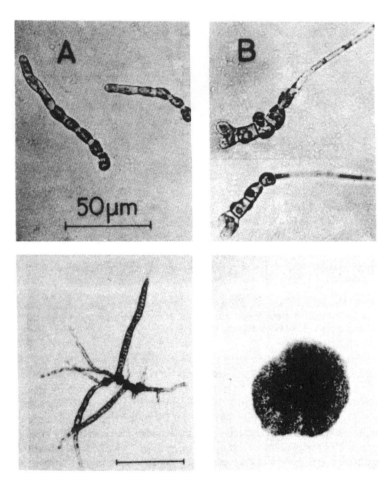

Fig. 5.11 Morphology of *Scytosiphon* after 9 days (*above*) and after 3 weeks (*below*) in red light (**A**) or in blue light (**B**) of equal photon irradiance at 15°C.[62]

details of the different types of spores and sporangia are omitted in order to concentrate on the biological function of different spore types in different algae. Additional details are given by Bold and Wynne,[19] and also by Chapman and Chapman.[33]

Types of reproduction in marine algae

In those marine algae which exhibit a single-phase life history (see Table 5.1), asexual reproduction is rare. Some of the macroscopic species may increase in numbers by vegetative propagation, but this usually involves the simple fragmentation of the thallus (e.g. *Sargassum*) rather than the production of any specialized structures, and reproductive development generally consists of gametogenesis alone.

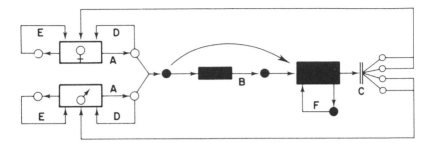

Fig. 5.12 Types of reproduction in marine algae with two or more morphological phases (haploid spores and phases shown white; diploid spores and phases shown black).
(1) Processes giving rise to another phase in the life history: **A**, gametogenesis; *B*, carposporogenesis (Florideophyceae only); **C**, meiosis followed by sporogenesis (e.g. in unilocular sporangia of brown algae, zoospores of green algae, tetraspores of Florideophyceae).
(2) Processes reproducing the same morphological phase: **D**, parthenogenesis of unfertilized gametes; **E**, neutral spores produced by haploid phase (e.g. monospores of gametophytes of red algae); **F**, neutral spores produced by diploid phase (e.g. in plurilocular sporangia of brown algae, monospores of sporophytes of red algae).

The production of gametes in all algae with complex life histories (i.e. two or more morphological phases; Table 5.1) is confined to the haploid, gametophyte phase (Fig. 5.12, A). In the more primitive brown and green algae (e.g. *Ectocarpus*, *Ulva*), any gametes which remain unfertilized are able to develop parthenogenetically into new gametophytes (Fig. 5.12, D), so that the gametes may also function as 'neutral spores', regenerating the same morphological phase. This process does not appear to occur in the more advanced brown algae (e.g. *Laminaria*) or in the red algae, which show a greater differentiation between male and female gametes. The gameto-phytes of many red algae, however, produce 'monospores', which develop into new gametophytes (Fig. 5.12, E) and so have a similar function to the unfertilized gametes of *Ectocarpus* and *Ulva*.

The diploid phases of algae with complex life histories can reproduce only asexually, but it is important to distinguish between processes which repro-duce the same morphological and cytological phase ('neutral spores', Fig. 5.12, F), and processes which give rise to another phase in the life his-tory (Fig. 5.12, B,C). Neutral spores produced by diploid thalli are always diploid, whereas (except in the carposporophytes of red algae; Fig. 5.12, B) the spores giving rise to a new phase are haploid, and their production involves meiosis (Fig. 5.12, C). Normally, the sporophytes of green algae form spores only by meiosis, so that self-propagation of the sporophytes is impossible. The same is true of the more advanced brown algae (e.g. all Laminariales), but the more primitive forms (e.g. *Ectocarpus*) produce diploid neutral spores by mitosis in many-celled (*'plurilocular'*) sporangia, as well as haploid spores by meiosis in single-celled (*'unilocular'*) sporangia. The two types of spores are indistinguishable — they are the typical reniform, biflagellate zoospores of brown algae — even though they are formed in different types of sporangia and develop into different types of thallus. The diploid conchocelis-phase of *Porphyra* (Table 5.1; p. 11) also produces neutral spores ('monospores') which regenerate the sporophyte, as well as haploid 'conchospores' which develop into the gametophyte, but the situation is less clear in the tetrasporophytes of the more advanced red algae. These plants typically produce tetraspores (so-called because they are formed in fours as the direct products of a meiotic division) which germinate into the gametophyte, but other types of sporangia are also observed in some species. Whether the spores produced in such sporangia function as tetraspores or neutral spores is not clear but, since the tetrasporophytes of some species can maintain themselves in areas that are unsuitable for the production or growth of the corresponding gametophyte (e.g. Trailliella-phase of *Bonnemaisonia*, see p. 146), such plants must be able to propagate themselves either through neutral spores or by vegetative reproduction.

Another aspect of the morphogenesis of marine algae that is best regarded as a type of reproductive development is the initiation of erect growth from a prostrate thallus. Many of the more delicate seaweeds growing in temperate waters are 'pseudo-perennials' or 'deciduous' forms ('deciduiphykes', see Table 7.2), in which the erect thallus dies down in unfavourable seasons (usually the summer) and the plant persists only as an inconspicuous crust or dwarf thallus. Since the erect thalli are more prolific spore producers and more effective spore dispersers than the prostrate thalli, the transition from prostrate to erect growth can be interpreted as the onset of reproduction in many species with different morphologies and different life histories.

Environmental control of reproduction

The long history of field observations of marine algae indicates that both the appearance of erect thalli in 'deciduous' species and the production of particular spore types in more persistent seaweeds occur at specific times of the year. Studies of algae in culture have confirmed that the transition from vegetative to reproductive development, or from one type of reproduction to another, can often be induced by a change in the environment. For example, sporophytes of *Ectocarpus* produce plurilocular sporangia at 19°C and unilocular sporangia at 10°C (Fig. 5.13). At intermediate temperatures, both types of sporangia may be formed, and the ratio between them is affected by other factors, such as irradiance.[167] These results indicate that the formation of gametophytes is favoured by low temperatures, and the transition from prostrate to erect growth also occurs in response to a decrease in temperature in several species (e.g. the green algae *Monostroma* and *Acrosiphonia*,[143] the brown alga *Desmotrichum*[211]). In some diatoms, on the other hand, sexual reproduction is induced by high temperatures, and resting spores, which are able to survive long periods of darkness in winter (e.g. up to 567 days at 0°C in *Thalassiosira*[67]), are formed when cultures are exposed to low temperatures (e.g. *Stephanopyxis*,[241] *Coscinodiscus*[257]).

In all of these responses, there seems to be a simple control of seasonal behaviour through temperature. Since temperature regimes in the sea are so much more stable than those on land (see p. 22), temperature provides a

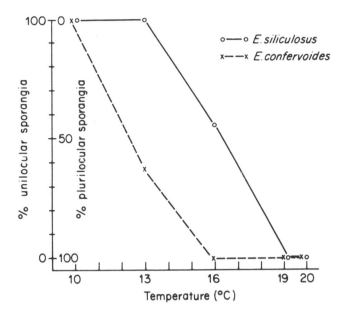

Fig. 5.13 Effect of temperature on the type of sporangia produced by two species of *Ectocarpus*.[167] The numbers of unilocular and plurilocular sporangia are expressed as a percentage of the total number of sporangia.

more reliable indication of the season for marine organisms than it does for terrestrial organisms. For this reason, it was long thought that the ability to measure and respond to daylength (***photoperiodism***), that is so often observed in terrestrial plants, would be of little ecological benefit to marine plants, and would not be observed. This idea has only recently been shown to be incorrect with the demonstration that the reproduction of several marine algae, representing all three major groups (Table 5.4), is stimulated by short days and that, as in all flowering plants, the response to short days is inhibited by a short period of light (a 'night-break') in the middle of a long night. The response of a plant to this wholly artificial light – dark regime is generally used as a diagnostic criterion for a genuine photoperiodic response in flowering plants,[249] but the application of the same criterion to algae assumes that the physiological mechanism of photoperiodism in algae is the same as in higher plants. The control of conchospore formation in the relatively primitive red alga *Porphyra* appears to justify this assumption, since the parallel with photoperiodism in flowering plants is very close.[59] For example, the photomorphogenetic pigment phytochrome appears to be responsible for light perception in the *Porphyra* response, even though there is no clear evidence for the action of phytochrome in any other non-green plant (i.e. any plant outside the Chlorophyta or higher plants). In the brown alga *Scytosiphon*, on the other hand, the initiation of erect thalli from the crustose *Ralfsia*-phase in response to short days (Table 5.4) is inhibited by very small doses of blue light, given as a night-break,[63] and this is the only

Table 5.4 Examples of marine algae whose reproduction is induced by short days (SD), but inhibited by a night-break (NB) in the middle of a long dark period.[59, 63, 143]

Species	Life history phase	Structure formed in SD	Most effective colour as NB
CHLOROPHYTA			
Monostroma grevillei	Codiolum (sporophyte)	Zoosporangia	?
PHAEOPHYTA			
Scytosiphon lomentaria	Ralfsia (sporophyte)	Erect thalli	Blue
RHODOPHYTA			
Porphyra tenera	Conchocelis (sporophyte)	Conchosporangia	Red
Bonnemaisonia hamifera	Trailliella (tetrasporophyte)	Tetrasporangia	?

photoperiodic response known in plants that does not appear to be controlled by phytochrome. Since the physiology of photoperiodism in *Porphyra* is so similar to that in flowering plants, it is perhaps surprising that a brown alga should show such a fundamental difference. However, the chemical structure of phytochrome is close to that of allophycocyanin (see p. 49), and this pigment is present in *Porphyra* but not in brown algae. These results suggest, therefore, that the different groups of marine algae may represent 'evolutionary experiments' in photoperiodism, as well as the 'experiments' in photosynthesis, thallus structure and life history that have already been noted (see p. 92).

In contrast to all of the responses discussed so far, gametogenesis in many species of Laminariales is independent of daylength, and is not specifically affected by temperature (i.e. reproduction occurs at all temperatures which support the growth of the gametophyte[143]). The critical environmental factor for the reproduction of laminarian gametophytes appears to be the presence of blue light. Female gametophytes grown in blue light undergo gametogenesis within 8 – 10 days of zoospore release, when the gametophyte consists of a single cell about 10 μm in diameter. In the same photon irradiance of red light, similar gametophytes continue to grow vegetatively for years — eventually forming balls of filaments up to 10 mm across — without producing a single female gamete, but 12 hours' irradiation of such plants with blue light is sufficient to induce gamete formation within 10 days.[145] Both the action spectrum and the photon requirements for this response are similar to those for the initation of two-dimensional growth in *Scytosiphon* sporelings (see p. 106), and this reproductive response may exert a similar control over the vertical distribution of laminarian sporophytes.[61]

Field observations of marine algae have clearly shown that some intertidal species may exhibit lunar rhythms of reproductive activity, as well as the annual or seasonal rhythms that are controlled by temperature or photoperiod. The commonest observation is that spore or gamete release occurs during each spring tide series, but does not occur during neap tides. This means that reproduction is confined to a 3–4 day period every 14 days. Periodic activity of this sort could be controlled by any of the environmental factors that are correlated with the transition from neap to spring tides (e.g. prolonged periods of emersion or submersion during neap tides, see p. 36), or it could be controlled by a natural rhythm — or 'biological clock' — within the plant itself. The existence of an endogenous lunar rhythm has been clearly demonstrated in the brown alga *Dictyota*, since female gametes continue to be produced at intervals of 14–15 days when the plants are isolated from all tidal influences and are maintained in laboratory culture (Fig. 5.14). The rhythm eventually dies away in laboratory conditions, but the clock can be 'reset' at any time by exposing the plants to low intensity light during the normal dark period. Gamete release then occurs 10 days later, and at 16–18 day intervals thereafter (Fig. 5.14). This effect of low irradiance during the night suggests that, in natural conditions, the plant's rhythm may be synchronized with the spring–neap tidal cycle by the light of the full moon (which coincides with spring tides, see p. 35) but, unfortunately, it is very difficult to prove that moonlight is really a significant factor in algal reproduction.

Endogenous control of sexual reproduction

The involvement of an endogenous rhythm in the gametogenesis of *Dictyota* indicates that algal reproduction is not entirely 'at the mercy of the elements',

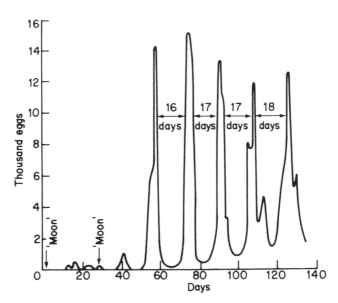

Fig. 5.14 Egg production by laboratory cultures of *Dictyota* growing in 14-hour days at 20°C.[167] The dark period was replaced by dim light on day 1 and day 28 to give the plants artificial 'moonlight'.

but that algae may exert some control over their own reproduction. This conclusion is supported by the ability of the female gametes of several brown algae to attract male gametes, and to stimulate their release from ripe antheridia, by secreting small amounts of organic chemicals into the surrounding water. Such chemicals are described as **pheromones**, rather than hormones, because they are secreted to the outside and they induce a change in the development or behaviour of other individuals of the same species, whereas hormones are involved in the internal control of growth and development of a single individual.

Pictures of the female gametes ('eggs') of *Fucus*, surrounded and almost smothered by the very much smaller male gametes, have long been used in text-books to illustrate sexual attraction, but it is only recently that sufficient attractant has been collected for chemical identification. Air was bubbled through suspensions containing a total of 3×10^{10} eggs (extracted from over 250 kg of *Fucus serratus* receptacles), and all volatile substances were condensed in a series of cold traps. Eventually, 690 µg of an oily material, smelling 'faintly flowery and linseed-like' was isolated, and christened 'fucoserraten'.[102] Surprisingly, this substance is not confined to *Fucus serratus*. The eggs of the closely related species *Fucus vesiculosus* secrete exactly the same pheromone,[170] and both species attract the male gametes of the other species as efficiently as they attract those of their own species.[171] Cross-fertilization rarely occurs between these two species, however, probably because macromolecules located on the surface of the gametes function as species-specific labels, and enable each gamete to recognize another of the same species.[20] Fucoserraten is an open-chain, unsaturated hydrocarbon (C_8H_{12}), but similar extraction techniques applied to the more primitive brown algae *Ectocarpus siliculosus* and *Cutleria multifida* have yielded slightly larger hydrocarbon molecules ($C_{11}H_{16}$) with a cyclic structure ('ectocarpen' and 'multifiden').[102] The male gametes of these two species also respond to the pheromone produced by the other species, but they are more sensitive to their 'own' pheromone.[169] Mature female gametophytes of several laminarian species have been shown to secrete a pheromone which triggers the release of male gametes from ripe antheridia, as well as attracting them to the female gametes. This pheromone has not yet been isolated or characterized, but all of the species investigated responded to the pheromone produced by all of the other species, so that a single substance appears to be secreted by all members of the Laminariales.[147]

In *Ectocarpus*, *Cutleria* and *Fucus*, the minimum concentration of pheromone that will induce a significant response in the corresponding male gametes is about 10^{-6}M, and the strength of the response increases with concentration above this minimum (Fig. 5.15). However, the effectiveness of ectocarpen increases less rapidly with concentration than does the effectiveness of multifiden or fucoserraten. The relative effectiveness of the different pheromones may be related to the pattern of sexuality in the different species. The female gametes of both *Ectocarpus* and *Cutleria* are motile flagellate cells, but those of *Ectocarpus* are similar in size to the male gametes (**'isogamy'**), whereas in *Cutleria* the females are about 35 times

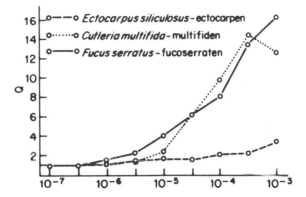

Fig. 5.15 Chemotactic response of male gametes of brown algae to the pheromone secreted by female gametes of the same species.[171] 'Q' represents the number of gametes attracted to the pheromone at any given concentration, divided by the number attracted to the pure solvent.

larger than the males (*'anisogamy'*). This means that a single female gamete of *Cutleria* represents a substantially greater reproductive investment than does a single female gamete of *Ectocarpus*, and it is not surprising that it should secrete a more effective pheromone. On the basis of this argument, fucoserraten should be even more effective, since the non-motile eggs of *Fucus* are some 20 000 times larger than the male gametes, but little additional increase in pheromone effectiveness seems to have accompanied the evolution from anisogamy in *Cutleria* to *oogamy* in *Fucus*. The characterization of pheromones from other brown algae may throw more light on their relationship to evolution in these plants.

Chemotactic attraction between gametes has also been observed in freshwater green algae and in aquatic fungi,[124] and similar mechanisms probably exist in all marine algae with flagellate male gametes (e.g. green algae, centric diatoms). It is perhaps less likely that red algae will exhibit pheromone activity, however, since the male gametes (*spermatia*) are non-flagellate, and are carried passively to the female *trichogyne* (a stigma-like extension of the female gametangium, which completely encloses a single female gamete). Nevertheless, one genus of red algae does show pheromone-like activity, although this is not related to normal sexual fusion, but to the fusion of somatic cells that occurs during the regeneration and replacement of dead cells in a filament. If a single cell dies in the uniaxial filaments of the large-celled alga *Griffithsia*, the cell above it regenerates a rhizoid from its base, and the cell below regenerates a 'repair shoot cell' from its apex. These two structures grow towards one another, and then fuse to replace the dead cell (Fig. 5.16a). A similar fusion can be induced between the cut ends of two separate filaments, provided that they are held close together during regeneration (e.g. in an empty cylinder of cell wall from the giant-celled alga *Nitella*), and that they are of the same species. If filaments from different species of *Griffithsia* are treated in this way, a normal apical cell (see p. 102) is regenerated from the cut end of the lower filament, rather than a repair shoot cell, and this does not fuse with the rhizoid produced by the upper filament (Fig. 5.16c). These observations suggest that the rhizoid regenerating from the upper filament secretes a pheromone which affects the type of regeneration from the apical end of the lower filament, and induces the two regenerating cells to grow towards one another. This pheromone, which has been called 'rhodomorphin',[251] must differ from species to species within the genus *Griffithsia*, and this indicates that some red algae can produce pheromones which — unlike those of the Phaeophyta — are species-specific.

GENETICS OF MARINE ALGAE

Marine algae have never attracted much attention from geneticists, probably because most species have complex life histories which, at best, take a long time to be completed in culture and, at worst, cannot be completed at all under controlled conditions. Another difficulty is that the physiology and morphology of these plants vary so much in response to changes in the environment that it is difficult to find a constant character whose inheritance can be studied. As culture methods improve, however, and as more is learnt about how to control the reproduction and the life histories of marine algae by manipulating their environment, so genetic investigations become more feasible and more productive.

Several genetical principles, long established on the basis of work with other organisms, can now be illustrated with examples from marine algae. Morphological variants in *Laminaria saccharina*[141] and in the red alga *Antithamnion*[212] are inherited according to standard Mendelian laws, whereas colour mutants in *Gracilaria* — a red alga which is rapidly becoming the *Neurospora* of the marine world — exhibit both Mendelian

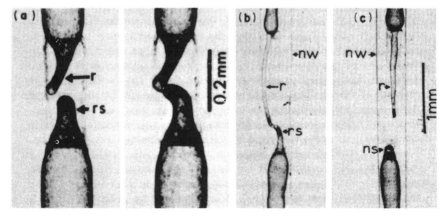

Fig. 5.16 Cell repair and somatic cell fusion in *Griffithsia*. **(a)** Normal repair mechanism following cell death within an intact filament of *G. pacifica*[253]; **(b)** regeneration of separate filaments of *G. pacifica*, resulting in cell fusion; **(c)** as **(b)**, but upper filament is *G. globulifera*, and no cell fusion occurs.[251] r = rhizoid regenerated from upper cell; rs = repair shoot cell; ns = normal shoot cell; nw = *Nitella* wall.

and cytoplasmic inheritance.[159] Several mutants of different colours — pink, yellow and green — have arisen spontaneously in cultures of this alga, while green mutants have been discovered in natural populations. These all provide good genetical 'markers' in crossing experiments, whose results suggest that the pigmentation of red algae is under the control of chloroplast DNA, as well as nuclear genes.[159]

Marine algae are particularly suitable organisms for studying the phenotypic expression of genes in the separate cytological and morphological phases of a single species. Almost all of our existing genetical knowledge is based on work with either diploid organisms, in which many recessive genes are hidden by their dominants, or haploid organisms, in which all genes are expressed but allelic interactions cannot be studied. Marine algae with two isomorphic generations (see Table 5.1) provide a unique compromise between these two situations, since the phenotypes expressed by two alleles separately in the haploid state can be compared with their combined effect on the same morphology in the diploid state. At present, even what determines whether a plant will develop as a gametophyte or a sporophyte is unknown. In the 'normal' situation, of course, haploid plants become gametophytes and diploid plants become sporophytes, but there are enough exceptions to this pattern to show that the level of ploidy, in itself, is not the critical factor in marine algae. The occurrence of haploid sporophytes and diploid gametophytes is well documented in the life histories of both *Ulva*[77] and *Ectocarpus*.[168] In these two genera, the morphology of the two phases is very similar but, in *Laminaria*, unfertilized haploid eggs frequently develop into small plants with the parenchymatous construction of the sporophyte, rather than the microscopic filaments of the gametophyte.[119] Among the red algae, tetrasporophytes of *Gracilaria* occasionally produce tetraspores without undergoing meiosis, and these develop into gametophytes even though they are diploid.[160]

There are also reports of so-called 'mixed phases' in several red algae, in which tetrasporangia and gametangia are formed on the same plant. While studying the inheritance of colour mutants in *Gracilaria*, van der Meer and Todd[160] obtained evidence for mitotic recombination, and suggested that such a phenomenon could explain the occasional appearance of male and female gametangia in the diploid tissue of a tetrasporophyte. If a pair of alleles, showing incomplete dominance, was responsible for the control of sex in red algae, the diploid tetrasporophyte cells would normally be heterozygous and, therefore, 'neuter', but recombination during a mitotic division would result in two daughter cells which were still diploid, but were now homozygous for either the male or the female gene (Fig. 5.17). Normal somatic growth following a single recombination of this type could produce two paired islands of male and female tissue within the sporophyte; this is the pattern of gametangia formation that is sometimes observed in the tetrasporophytes of *Gracilaria*.

This suggestion that sex in red algae is controlled by a pair of alleles, rather than by the morphologically distinct X- and Y-chromosomes of most other organisms (including Laminariales[74]) is perhaps the first really startling

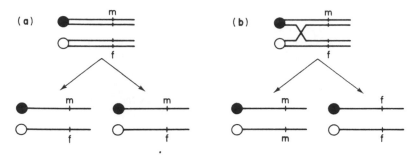

Fig. 5.17 Genetic mechanism by which sexual tissue could arise in diploid tetrasporophytes of red algae.[160] **(a)** Normal mitotic division; **(b)** mitotic recombination resulting in diploid cells which are homozygous for the sexual genes.

result to emerge from genetic studies of marine algae, and it may help to persuade geneticists to distribute their efforts more evenly among algal genera. The contents page of a recent monograph on *The Genetics of Algae*[134] reads like a score-line:

Chlamydomonas: 144 pages   Multicellular marine algae: 17 pages.

6

The Ecology of Seaweeds: Zonation and Succession

Seaweeds are not simply plants that live in the sea: they are morphologically and physiologically distinct from both land plants and the majority of fresh-water algae (see Chapter 1). The detailed study of the relationships between these plants and their natural environment may, therefore, provide some clues about why seaweeds are so different from the plants of other habitats. Rocky shores and solid substrates down to the lower limit of the photic zone provide the main habitat for seaweeds, and this habitat, in itself, is of interest to plant ecologists because of the large variations in the degree of distur-bance and stress that may occur within relatively small areas. *Disturbance* is used here to refer to factors, such as wave action and grazing, which limit plant biomass by causing its partial or total destruction, whereas *stress* is used as a collective term for the external factors which limit the rate of dry matter production by plants.[85] It is excessive disturbance which prevents seaweeds from colonizing mud, sand or shingle, and which restricts their development on wave-beaten rocks, but the limits to the vertical distribution of seaweeds on sheltered rocky shores are largely imposed by the stresses associated with emersion at the top of the shore, and by low light at the bottom of the photic zone. Between these two extremes, and especially in the upper part of the subtidal zone, the intensities of both stress and disturbance may be low, and conditions are then particularly favourable for plant growth (see p. 85). However, this results in increased *competition* between plants for light and space, and the competitive ability of plants becomes more important in these habitats than their ability to withstand either stress or disturbance. The patterns of vertical distribution and of succession in seaweed communities are best studied with reference to local variations in the intensities of disturbance, stress and competition, particularly as this approach has proved fruitful when applied to the ecology of terrestrial plants.[85]

GENERAL FEATURES OF BIOLOGICAL ZONATION ON ROCKY SHORES

One of the most conspicuous and well-known characteristics of the biology of rocky shores is the horizontal banding of the intertidal zone that is caused by the dominance of different organisms at different heights on the shore. This **zonation** occurs on shores throughout the world, and the organisms involved include encrusting lichens (e.g. the black lichen, *Verrucaria*) and sedentary animals, such as barnacles and mussels, as well as different species of seaweeds. The boundaries between different organisms are often so sharp and so level that it is tempting to conclude that there is a direct causal relationship between such boundaries and the water level at particular states of the tide. The apparent levelness of these boundaries is often deceptive, however, and the absolute height of a specific boundary may change significantly within a short distance on a single shore. For example, a change in the slope or the aspect of the shore will alter the degree of exposure to wave action, and the upper limits of most species are found to be higher in more exposed sites (Fig. 6.1). Other factors which reduce the severity of thermal and desiccation stress during emersion (see p. 37) have a similar effect. It is clear, therefore, that the tidal levels themselves do not determine the heights of the biological boundaries, and it is not possible to describe the zonation, or to define the boundaries between zones, in terms of critical tidal levels such as EHWS or E(L)HWN (see p. 36).

For this reason, most recent investigators have described the distribution of species on their shores with reference to three biological boundaries, all of which can be identified on rocky shores at most latitudes in every ocean.[238] From the top of the shore, these boundaries are:

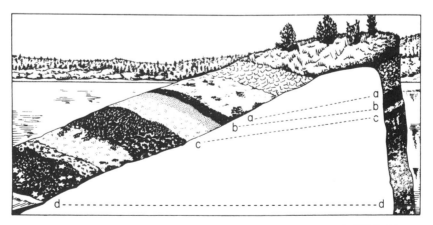

Fig. 6.1 Diagrammatic section through Brandon Island, British Columbia.[238] The island is located in a sheltered bay on the east coast of Vancouver Island and neither shore experiences severe wave action, but the gentler slope faces south. **a – a** : upper limit of *Verrucaria*; **b – b**: upper limit of barnacles; **c – c**: upper limit of main *Fucus* population; **d – d**: low water of a low spring tide.

1. the upper limit of *Verrucaria*, blue-green algae and littorinid snails;
2. the upper limit of barnacles;
3. the upper limit of laminarians.

Although there is general agreement on the validity of these boundaries, there is less agreement on what to call the zones between. The terminology adopted here follows that suggested by Lewis[135] and is illustrated in Fig. 6.2, together with the general relationship between these biological boundaries and tidal levels under different degrees of exposure. The *littoral fringe* may be contained entirely within the intertidal zone in sheltered situations, but the whole of the littoral fringe and most of the *eulittoral zone* may be above the level of the highest tides under conditions of extreme exposure (Fig. 6.2). The upper part of the *sublittoral zone*, lying between ELWS and the upper limit of the laminarians, is sometimes described as a separate 'sublittoral fringe',[238] but the transition from this fringe to the rest of the sublittoral zone is not always distinct, and it is not possible to define a biological boundary at this level which can be applied as widely as the three major boundaries.

The species or groups of species whose upper limits are used to define the boundaries between the various zones have been selected because they are found in most latitudes and on shores subjected to a wide range of exposure. Most of the other species on rocky shores show a more restricted distribution with respect to both latitude (see Chapter 7) and exposure. For example, the large brown algae (*Fucus* and related genera), which dominate the eulittoral zone of sheltered shores in most temperate regions of the world, gradually

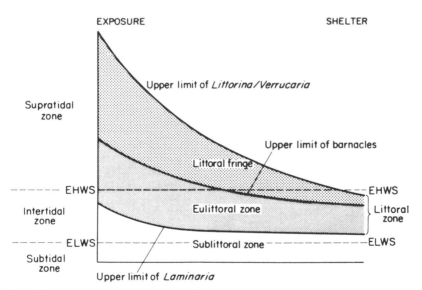

Fig. 6.2 Generalized scheme of biological zonation on rocky shores, and the relationship between biological zones and tidal levels under different intensities of exposure.[135]

decrease in abundance as wave action increases, and are entirely replaced by barnacles and mussels on very exposed shores (see p. 136). Exposure also affects the species composition of the upper subiittoral zone. *Alaria* is the characteristic kelp of this zone on exposed coasts in the northern hemisphere, whereas *Laminaria saccharina* occurs at the same level on more sheltered shores. Further details about the species composition or the vegetation of rocky shores are given by Stephenson and Stephenson[238] and by Lewis.[135]

QUANTITATIVE ECOLOGICAL STUDIES OF ZONATION

This biological classification of rocky shores has facilitated the qualitative comparison of zonation patterns on widely different shores, and has established certain generalizations about zonation that are valid throughout the world. There is always the danger, however, that attempting to classify the shore on the basis of the presence or absence of a limited number of species will lead into circular arguments, similar to those discussed in connection with biological exposure scales (e.g. 'species A, B and C always occur in different zones', but these zones have been separated only because of the presence of A, B and C, respectively; see also p. 42). In detailed studies of particular shores, it is important to establish that the upper limits of the key species really do coincide with the transition from one community to another, and are not just arbitrary lines drawn across a map of the shore.

The objective, quantitative methods developed by terrestrial plant ecologists[121] can readily be applied to littoral vegetation and, since many of the animals are attached forms or are quiescent during emersion, the fauna can be included as well. In the example shown in Fig. 6.3, two quadrats of 0.25 m² were examined at each of five sites at 1 m vertical intervals down a rocky shore in the Isle of Man, and the species present in every 10 × 10 cm

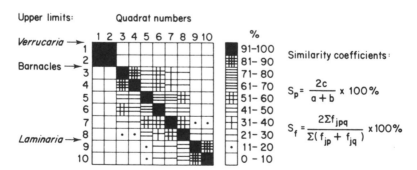

Fig. 6.3 Similarity matrix for 10 quadrats (0.25 m²) at different heights on a rocky shore in the Isle of Man (adapted from Russell[215]). (*Below diagonal*) Similarity coefficient based on presence and absence data (S_p), where a, b = number of species in first and second quadrats, respectively, and c = number common to both quadrats. (*Above diagonal*) Similarity coefficient based on frequency data (S_f), where f_{jp}, f_{jq} = frequency of species j in first and second quadrats, respectively, and f_{jpq} = *lesser* of the two values f_{jp} and f_{jq}.

subdivision of each quadrat were recorded.[215] Two simple similarity coefficients were calculated for every pair of quadrats, one based on the overall species list for each quadrat, and the other on the frequency with which each species occurred in the subdivisions. Quadrats 1 and 2 were located in the littoral fringe and were very similar, but none of the 8 species in these quadrats was found in any other quadrat. Thus, the littoral fringe was quite distinct from the eulittoral zone in this site, and the upper limit of barnacles provided a good indicator of the boundary between these zones (Fig. 6.3). The boundary between the eulittoral and the sublittoral zones was less well-defined, however, since quadrat 8 shared nearly 50% of its species with quadrats 9 and 10, even though it was above the upper limit of *Laminaria*. When the similarity coefficient based on the frequency of the different species was used, the boundary became slightly clearer, but it was still not as sharp as that between the eulittoral zone and the littoral fringe. The low similarities between quadrats 3 and 4 and quadrats 7 and 8 (Fig. 6.3) indicate that the top and the bottom of the eulittoral zone are quite different, but the transition is too gradual to permit any subdivision of this zone. In a similar study at another site, Russell[214] was unable to distinguish a sublittoral zone at all, and the boundary between the eulittoral zone and the littoral fringe was less well-marked than in Fig. 6.3.

The few studies of this type that are available clearly indicate the limitations of the three-zone concept of rocky shores, and illustrate the potential of the quantitative approach. Our understanding of the complexities of rocky shore ecology would benefit from a wider application of these methods, particularly in similar sites with different exposure. The qualitative effects of exposure on the upper limits of conspicuous species are well known (Fig. 6.2), but it is by no means certain that all species react to exposure in the same way. Unfortunately, the lower limits of quantitative ecologists are also likely to be significantly affected by the degree of exposure but, hopefully, their scientific curiosity will sometimes be stronger than their instincts for mere self-preservation!

FACTORS CONTROLLING THE VERTICAL DISTRIBUTION OF SEAWEED SPECIES

The sharp horizontal limits to the distribution of many species on rocky shores were investigated many times during the first 50 years of this century in relation to the so-called 'tide-factor hypothesis', which suggested that the length of emersion was critical in determining the height on the shore at which different species could live. Many of these investigations failed, however, to take account of the variation of emersion time at a fixed point on the shore during the spring – neap cycle of the tides, or to quantify the effects of exposure on both emersion times and distribution limits. More and more attention was concentrated on experiments testing the ability of different species to survive periods of drying at high temperatures or abnormal salinities, which bore little resemblance to the conditions experienced by the plants in their natural environments. Apparent correlations between the resistance of species to such treatments and their position on the shore have often been reported, but the ecological irrelevance of many of the experiments led Chapman[29] to conclude that 'the search for causes of zonal discontinuities has been fruitless . . . . . because ecological zonation is largely dependent on the competitive relations of species . . . . . and physiological tolerance limits play no part in this explanation' (p. 69). This was possibly intended as a deliberate overstatement of the case against physical factors and physiological tolerances in the control of littoral zonation, and it certainly seemed to stimulate a revival of interest in the problem and a greater concern for the ecological realism of experimental treatments. This section, therefore, concentrates on a few recent studies of the physiological ecology of rocky shore algae.

Much of this work concerns a group of five, closely-related brown algae (all members of the Fucales, or 'fucoids'), which occur commonly and in a wholly predictable vertical sequence on sheltered shores in N.W. Europe (Fig. 6.4). The best evidence for the importance of physical factors in the zonation of these species comes from observations of natural populations on the shore throughout the year.[221] On several occasions, the uppermost plants

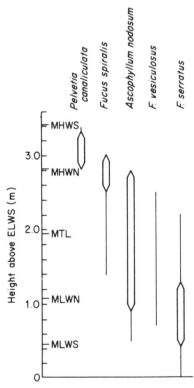

Fig. 6.4   Vertical distribution of five common intertidal fucoid algae on a sheltered shore in western Scotland, compared with tidal levels.[221]

of *Pelvetia*, *Fucus spiralis* and *Ascophyllum* showed distinct tissue damage 3 – 4 weeks after dry, sunny weather had coincided with neap tides (when emersion times are longer at the top of the shore, see p. 36) during the spring and summer, and this was followed by the death of these plants and a consequent pruning back of the upper limit of each species. Neap tides during the winter did not have the same effect, regardless of the weather conditions, so that prolonged exposure to low temperatures or to rain was not as damaging as high temperatures and severe desiccation. The upper limits of *F. vesiculosus* and *F. serratus* were not affected by severe drying conditions during neap tides, but transplantation of these species to higher levels on the shore caused a significant reduction in growth rate, whereas the growth of *F. spiralis* and *Ascophyllum* was similar at all positions in the eulittoral zone (Table 6.1). The results of this and other transplantation experiments[221, 225] suggest that the upper limits of fucoid algae are determined by the physiological stresses associated with emersion, but that the lower limits may be imposed by competition with the species occupying the next zone on the shore. These hypotheses will now be examined in more detail.

Upper limits: stresses associated with emersion

The most obvious of these stresses is desiccation and this has received the most attention, both in early work and more recently. The first question to ask is: Do upper shore algae lose water more slowly than lower shore algae?

Rates of drying

Some of the earlier investigators reported an inverse correlation between the rate of water loss in various seaweeds and their height on the shore, but the differences between species were mostly very small and were probably not statistically significant. Recent studies of European fucoids[129] and a range of Australian algae from intertidal and subtidal habitats[64] have shown that the rate of water loss is largely dependent on the surface area:volume ratio of the tissue, and bears no consistent relationship to the natural habitat of the

Table 6.1 Relative growth in fresh weight (g g^{-1} day^{-1}) of four species of fucoid algae transplanted into different regions of the eulittoral zone. The experiment was conducted in Co. Down, Northern Ireland, over 2 – 3 weeks in September 1979 (unpublished data of F.A. Brown).

| Species | Region of eulittoral zone | | |
	Top	Medium	Bottom
Fucus spiralis	0.015$_a$	0.017$_a$	0.019$_a$
Ascophyllum nodosum	0.004$_a$	0.006$_a$	0.007$_a$
Fucus vesiculosus	0.016$_a$	0.015$_a$	0.024$_b$
Fucus serratus	0.002$_a$	0.016$_b$	0.023$_c$

For each species, values with different subscripts are significantly different at $P = 0.05$

species. Even factors such as cell wall thickness and mucilage content were found to be unrelated to the rate of drying. Schonbeck and Norton[222] examined several possible ways in which water loss might be reduced in intertidal fucoids, and came to the conclusion that avoidance of desiccation by reducing water loss was not the primary adaptive mechanism in these species. In the absence of adaptations equivalent to the thickened cuticles and sunken stomata of xerophytic flowering plants, desiccation-tolerant algae might be expected to show physiological adaptations which either reduce the effects of drying on the metabolism of the plant, or ensure the rapid recovery of metabolic processes after re-immersion.

Effects of drying on photosynthesis

The photosynthetic rate of a seaweed in air is usually estimated from CO_2 measurements using an infra-red gas analyser (IRGA), since the oxygen technique (see p. 86), which would be used for the same plant in water, is clearly inappropriate. It is difficult, therefore, to obtain a direct and reliable comparison of the submersed and emersed rates of photosynthesis. However, the photosynthetic rate of several species has been found to increase as the thallus dries after emersion, and to reach a peak when the alga has lost $10-20\%$ of its initial water content.[109, 201] This effect may be related to an initial increase in the rate of diffusion of CO_2 into the plant as the surface water dries, but it is soon followed by a steady decline in photosynthetic rate as water loss continues. An investigation of the changes in chlorophyll fluorescence from seaweed thalli with different water contents[263] indicated that the first deleterious effect of drying was the uncoupling of electron transport between photosystems I and II but that, when the thalli had lost more than $60-75\%$ of their water, electron transfer from the photolysis of water to photosystem II stopped and photosynthesis ceased. The precise water content at which this occurred was found to vary from species to species, but no correlation was apparent between the critical water content and the resistance of the species to desiccation. The relationship between the decline in photosynthesis and the water content of the thalli was also very similar in three species of *Fucus* (Fig. 6.5), and there was no indication that the photosynthetic apparatus of *F. spiralis* was more resistant to water loss than that of *F. serratus*. The physiological mechanism of desiccation tolerance must, therefore, involve the rate of recovery after drying, or the extent of this recovery.

Recovery of photosynthesis after drying

The rate of recovery of photosynthesis after a period of drying is clearly important in upper shore algae, which may be covered by water for only $1-2$ hours in each tidal cycle, but little difference could be detected between the speed of recovery in the littoral fringe species *Pelvetia canaliculata* and that of the sublittoral *Laminaria digitata*. The photosynthesis of several

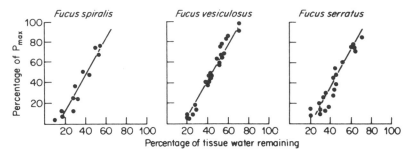

Fig. 6.5 Photosynthetic rate of fucoid algae during emersion in relation to the water content of the tissues. Photosynthesis is expressed as a percentage of the maximum rate for each plant, which was usually recorded when about 80% of the initial water remained (unpublished data of F.A. Brown).

fucoids and *L. digitata* increased to a steady rate within 15–30 minutes of re-immersion after moderate drying, but recovery took nearly 2 hours in all species if the drying treatment was severe. There were marked differences between the species, however, in the extent of the recovery after drying (Fig. 6.6). The steady rate achieved by *L. digitata* on re-immersion was not as high as the maximum rate before drying if the thalli had lost more than 40% of their initial water content and, if the water loss was more than 80%, no photosynthesis could be detected at all after re-immersion. In *Pelvetia*, on the other hand, the rate of photosynthesis after re-immersion was similar to that before drying even in thalli that had lost all but 4% of their initial water content. The extent of the recovery of photosynthesis in the three species of *Fucus* (Fig. 6.6) was intermediate between that of *Pelvetia* and *L. digitata*, and correlated well with their natural zonation. Plants of *F. vesiculosus* from the upper limits of this species also showed better recovery of photosynthesis than plants of the same species from lower on the shore.

Such intra-specific differences in desiccation tolerance have also been observed in *Fucus spiralis*.[223] Individuals from the upper limits of this species showed better growth after a severe drying treatment than plants from the bottom of the *F. spiralis* zone, and plants collected during the summer appeared to be more tolerant of desiccation than those collected from the same position during the winter. This last observation suggests that intra-specific variations result from phenotypic adjustments in physiology, equivalent to the drought-hardening of flowering plants,[132] and not to the natural selection of genetically adapted individuals. This hypothesis was supported by an experiment in which increased desiccation tolerance was induced in young plants of *F. spiralis* by exposing them to air for 2–4 hours each day for 5 days.[223] The growth of these 'hardened' plants was not significantly affected by a 44-hour drying treatment, whereas the growth of unhardened plants after the same treatment was reduced by about 75% (Table 6.2).

The primary criterion of desiccation tolerance in these experiments was the growth rate of the plants following a severe drying treatment, but two other characteristics of the plants showed a good correlation with the growth

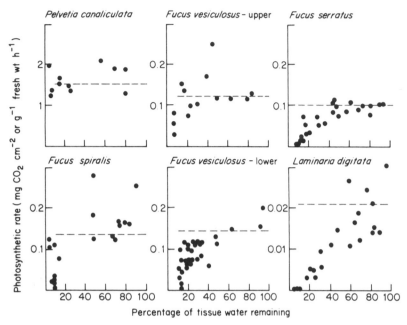

Fig. 6.6 Maximum rates of photosynthesis of brown algae recorded in air after recovery from drying treatments which had reduced the tissue water by different amounts (unpublished data of F.A. Brown). The mean value for the maximum rates recorded for each species before drying is indicated by a broken horizontal line. The photosynthetic rate of *Pelvetia* is expressed per unit fresh weight, and that of all other species per unit area.

criterion. These were the percentage of dry matter in the tissues, and the percentage change in fresh weight of the tissue following drying and re-immersion (Table 6.2). These two characteristics are very much easier to measure than rates of photosynthesis or growth after drying, and their validity as estimates of the desiccation tolerance of different seaweeds needs to be thoroughly tested on a wide range of species, subjected to a variety of growth conditions. If the correlations are confirmed, an examination of the underlying causal relationships will clearly be the next step in unravelling the mechanism of desiccation tolerance in seaweeds.

Nutrient stress during emersion

When *Fucus serratus* was cultured in a simulated tidal regime with 11 hours' emersion every 12 hours, the growth rate was 40% lower than in regimes with only 1 or 6 hours' emersion.[224] This result would not be surprising if it were not that the plants were maintained at a high humidity during emersion, and lost very little water. The photosynthesis of emersed fucoids has been found to remain high under these conditions, so that the reduction in growth during prolonged emersion cannot be attributed either to desic-

Table 6.2   Desiccation tolerance of hardened and unhardened plants of *Fucus spiralis*, measured by linear growth in culture after an additional drying stress, and correlated with the dry matter content of the tissues and with the change in fresh weight following drying and re-immersion. The hardening treatment consisted of drying the plants for 2 – 4 h at 25 – 27°C and 50 – 52% relative humidity on each of 5 successive days.[223]

	Percentage dry matter	Additional drying stress	Response after re-immersion	
			Change in fresh weight	Linear growth (mm 25 days^{-1})
Hardened	25.0	44 h	− 2.4%	14.2
		None	–	16.2
Unhardened	20.9	44 h	− 13.5%	4.6
		None	–	19.5

cation stress or to reduced photosynthesis. One hypothesis which could explain the effect of emersion on growth is that the plants were unable to obtain sufficient nutrients during short periods of submersion (see p. 38), and this hypothesis has been tested by conducting a similar experiment with *Pelvetia* and *F. spiralis* in artificial tidal regimes, using sea water enriched with different concentrations of nutrients (Table 6.3). In unenriched sea water, both species grew more slowly when emersed for 10 hours than when emersed for 2 hours, although the difference between the two tidal regimes was substantially greater for *F. spiralis* than for *Pelvetia*. The effects of adding nutrients to the sea water was the same in both species: growth increased in the 10-hour emersion regime, but showed little overall change in the 2-hour regime (Table 6.3). These results provide clear support for the idea that nutrient stress may be just as important as desiccation stress in the intertidal zone, and suggest that one reason why *Pelvetia* is the 'top fucoid' on sheltered shores may be its ability to survive on lower levels of nutrients than *F. spiralis*. Whether this ability is due to more rapid nutrient uptake during submersion, or to lower nutrient requirements (related to the lower growth rate) has yet to be established.

Table 6.3   Linear growth (mm month^{-1}) of two species of fucoid algae cultured for 6 – 8 weeks in two artificial tidal regimes and in sea water with different concentrations of added nutrients.[224]

	Pelvetia canaliculata		*Fucus spiralis*	
Nutrient enrichment	Tidal regime (h emersion per 12 h cycle)			
	10 h	2 h	10 h	2 h
Full strength	5.6$_a$	7.8$_a$	15.1$_a$	12.2$_b$
1/5 strength	4.9$_a$	5.1$_b$	4.5$_b$	20.7$_a$
1/20 strength	2.8$_b$	7.2$_a$	4.5$_b$	21.6$_a$
Unenriched	1.7$_b$	5.3$_b$	1.5$_c$	10.9$_b$

In each column, values with different subscripts are significantly different at P = 0.05

Lower limits: competition, irradiance and grazing

Some of the experiments already described have shown that the growth of the upper shore fucoids is not inhibited when they are transplanted to positions lower on the shore (Table 6.1) or cultured in tidal regimes typical of the lower intertidal zone (Table 6.3). Indeed, transplantation to mid-shore positions has been observed to stimulate the growth of *F. spiralis* and *Pelvetia*, provided that the transplants were located in areas cleared of *Ascophyllum* and *F. vesiculosus* — the species which occur naturally in that zone.[225] These results indicate that the physical conditions below the normal lower limits of *Pelvetia* and *F. spiralis* are not unfavourable to the growth of these species, and that their lower limits must, therefore, be imposed by competition with the fucoids of the middle and lower shore.

The importance of competition has been demonstrated by following the natural re-colonization of areas of the shore which have been cleared of their existing flora. For example, a patch of rock (0.5 × 1.0 m) within the *F. spiralis* zone, which was cleared of all macroalgae and barnacles in the summer, was found to contain 2 – 3 microscopic fucoid germlings per mm² after 8 months.[225] Half of these germlings were too small to be identified with certainty at this stage, but a substantial proportion of the remainder (14%) were *Pelvetia* and the rest were *F. spiralis*. However, the relatively slow-growing plants of *Pelvetia* (see Table 6.3) were rapidly overshadowed by a closed canopy of *F. spiralis*, and the density of *Pelvetia* gradually decreased until the species disappeared completely. In plots in which the plants of *F. spiralis* were continually removed as soon as they were large enough to identify, the *Pelvetia* developed normally and soon formed a conspicuous patch within the *F. spiralis* zone. Near the lower limit of *F. spiralis*, young plants of *Ascophyllum* were found to show continued growth underneath the *F. spiralis* canopy, and the dominance of *Ascophyllum* over *Fucus* species in the middle of the shore is probably achieved, in spite of its slower growth rate (Table 6.1), by this tolerance of shaded conditions, and the longevity of individual plants (15 – 25 years[51]). Thus, *F. spiralis* and *F. vesiculosus* may overshadow *Ascophyllum* in the early stages of re-colonization but, when the first generation of *Fucus* plants dies after 3 – 4 years, the next generation cannot become established beneath the persistent understorey of *Ascophyllum*. Selective grazing may also be important in the competition between fucoid genera. Littorinid snails show a distinct preference for *Pelvetia* and *Fucus* germlings, and have been reported to refuse *Ascophyllum* even after 12 weeks without any other food.[225]

Most of the competition between different species of fucoid algae on the upper and middle shore must be for light, since all of the plants at any one tidal level receive the same nutrient supply, and there is so much bare rock between the holdfasts of individual plants that there can be little competition for primary space. Competition for light also seems to be important on the lower shore and in the upper sublittoral zone. In the same way that fast-growing *F. spiralis* overshadows *Pelvetia*, and the long adult fronds of *Ascophyllum* overshadow *Fucus* species, especially when buoyed up by their air-bladders during submersion, the canopy formed by the upper sublittoral laminarians will dominate and impose a lower limit on *F. serratus*. These uppermost laminarians, all of which possess flexible stipes (e.g. *Laminaria digitata, Alaria*) will, in turn, be overshadowed by species with rigid stipes (*L. hyperborea, L. ochroleuca*) in habitats that are more than a stipe's length below the lowest low waters. In N.W. Europe, *L. digitata* rapidly colonized areas that were cleared within the sublittoral forest of *L. hyperborea*, but was gradually eliminated as the canopy was re-established.[119] On the other side of the Atlantic, however, *L. hyperborea* is absent, and *L. digitata* is abundant down to 15 – 20 m.

Competition cannot account for the lower limit of the laminarians, as a group, since the seaweeds that are found at greater depths are all low-growing, delicate forms, or crustose and often calcified species. Similarly, the maximum depth of these deep sublittoral algae cannot be determined by competition, since there is only bare rock below them. The depths at which both of these lower limits occur increase with the clarity of the water (see p. 18; Table 2.1), and the light available for photosynthesis is almost certainly the critical factor (see p. 53). It appears, therefore, that the light requirements of a species, or its tolerance of low irradiances, are the ultimate determinants of the lower limits of most species, regardless of whether these limits are actually imposed by the physical environment or by competition with another species. The light compensation point for photosynthesis is probably the best indicator of ability to grow at low irradiances but, because of the phenotypic adaptation of plants to different temperatures and growth irradiances (see pp. 55, 79), few of the available estimates of compensation points in different seaweeds are really comparable. However, the irradiance required to saturate photosynthesis does show a clear relationship with the ecological zone that a species occupies (Table 6.4). Intertidal seaweeds from all three groups are saturated at photon irradiances of about 500 μmol m^{-2} s^{-1}, whereas the laminarians and red algae of the upper sublittoral zone require only 150 – 200 μmol m^{-2} s^{-1} for saturation. Finally, the red algae that are typical of deep-water habitats are saturated at 60 – 70 μmol m^{-2} s^{-1}.

In a few sites, the lower limit of an algal species is not controlled by irradiance or by competition with another seaweed, but depends upon the activities of herbivorous animals. The clearest examples are the effects of sea urchins on the spread of sublittoral kelp populations. The monthly removal of *Echinus* from an area immediately below the *Laminaria hyperborea* forest off the Isle of Man caused a significant downward extension of the forest within 2 – 3 years,[111] and a central Californian kelp forest also expanded rapidly following the mass mortality, through natural disease, of a population of *Strongylocentrotus*.[187] It is possible that the stress caused by low irradiance in these sublittoral habitats contributes to the sensitivity of such algal communities to grazing pressure. Grazing animals in the intertidal zone, such as limpets and littorinid snails, are also important, but they usually affect the species composition over the whole area (see p. 136), rather than the position of the vertical limit for a particular species (but see below).

Table 6.4 Photon irradiance required for saturation of photosynthesis in benthic marine algae from different vertical zones.[144]

Zone	Species	Photon irradiance $(\mu mol\ m^{-2}\ s^{-1})$
EULITTORAL	Fucus vesiculosus (B)	600
	Fucus serratus (B)	500
	Codium fragile (G)	500
	Monostroma nitidum (G)	500
	Gigartina stellata (R)	460
	Porphyra tenera (R)	400
UPPER SUBLITTORAL	Laminaria spp. (European; B)	150
	Laminaria japonica (B)	200
	Macrocystis pyrifera (B)	220
	Undaria pinnatifida (B)	200
	Chondrus crispus (R)	180
	Palmaria palmata (R)	210
	Phyllophora antarctica (R)	160
DEEP SUBLITTORAL	Delesseria sanguinea (R)	60
	Ptilota serrata (R)	70
	Plocamium telfairiae (R)	60

G = green algae; B = brown algae; R = red algae

Zonation of seaweeds: conclusions

There is good evidence from field experiments that the lower limits to the vertical distribution of many seaweed species in the littoral and upper sublittoral zones are imposed by competition for light with other seaweed species, but there is, as yet, no evidence that upper limits may be imposed in the same way. However, a recent large-scale field experiment in S.E. England, which was conducted unintentionally by the wreck of the oil-tanker *Torrey Canyon*, has shown that the upper limits of some sublittoral algae may be imposed by grazing.[235] The massive quantities of dispersants which were used to remove the oil that came ashore are thought to have caused more ecological damage than the oil itself, and resulted in the death of most of the intertidal herbivores. Wave-beaten shores, which had been dominated by barnacles and limpets, were rapidly colonized by the green alga *Enteromorpha* and then by *Fucus*, and the upper limits of *Laminaria digitata* and the sublittoral fucoid *Himanthalia* were raised by up to 2 m (Fig. 6.7a). The gradual recovery of the limpet population was followed, 5 – 10 years after the original disturbance, by the re-establishment of a barnacle-dominated eulittoral zone, and the return of the eulittoral/sublittoral boundary to its original height (Fig. 6.7b).

Thus, the temporary absence of herbivores resulted in a rise in the upper limits of these sublittoral algae — but only to a new level that was presumably imposed by the stresses associated with emersion, and it seems probable that the upper limits of most seaweeds are ultimately determined by such

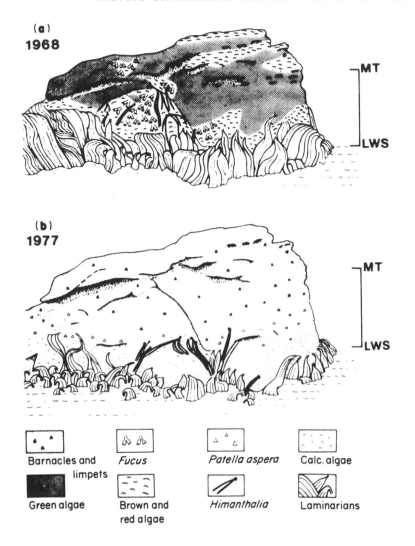

(a) 1968

MT

LWS

(b) 1977

MT

LWS

Barnacles and limpets

Fucus

Patella aspera

Calc. algae

Green algae

Brown and red algae

Himanthalia

Laminarians

Fig. 6.7 Wave-beaten outer rocks near Cape Cornwall, England, **(a)** 13 months and **(b)** 10 years after the oil spill from the *Torrey Canyon*[235]. Upper limits of *Laminaria digitata* and *Himanthalia* were 1.5 to 2 m higher in 1968 than in 1977.

stresses in their abiotic environment. Recent laboratory experiments have reduced the wide range of possible explanations for these effects (see pp. 125–9), but the *exact* physiological cause for a specific limit, or the *precise* environmental factor controlling it in the field, has not yet been established for any species at any site. This is a reflection, however, of the complexity of the problem, rather than of the competence of the scientists who have tackled it!

SUCCESSION IN SEAWEED COMMUNITIES

The process of ecological succession occurs more rapidly in benthic marine habitats than in terrestrial communities, largely because the life-spans of the dominant organisms involved are shorter in the sea than on land. Few seaweeds are known to live as long as *Ascophyllum nodosum*, but the maximum life-span of this species (possibly 25 years) is far shorter than that of most forest trees. Benthic marine communities, therefore, provide a particularly convenient testing-ground for theories about ecosystem development and ecological succession, and they have received increasing attention from ecologists in recent years.

This attention has generally taken the form of field experiments in which the pattern of re-colonization has been studied following one of three types of artificial disturbance. The commonest approach has been to scrape the rock surface clean of all macroscopic plants and animals, and sometimes to follow this with chemical treatment to kill microscopic organisms and spores, but alternative approaches, which bear a closer resemblance to occasional natural disasters, have involved the selective removal of either herbivores (e.g. limpets, sea urchins) or the seaweed canopy. On the basis of experiments of this last type, conducted on the coast of Washington, Dayton[53] has described three ecological categories of algae.

(1) *Canopy species*: growing above other species and dominating the light resource — mainly kelps;

(2) *Obligate understorey species*: growing below the canopy and decreasing in abundance, through desiccation, excessive light or wave action, when the canopy is removed — mainly red algae;

(3) *Fugitive species*: colonizing rapidly in response to any disturbance — includes green, brown and red algae.

The characteristics of the latter two categories are well illustrated by the results of one experiment (Fig. 6.8), in which the canopy-forming kelp *Hedophyllum* was removed from an area of 4 m². The exclusion of the herbivorous sea urchin *Strongylocentrotus* from strips of shore in the same locality also caused an immediate bloom of fugitive species (especially *Ulva*, *Enteromorpha* and *Porphyra*) in areas which had previously been devoid of seaweed cover. This bloom was followed by the gradual establishment of a *Hedophyllum* canopy and an associated flora of obligate understorey species.[53] A similar pattern of succession was observed in S.W. England when the intertidal limpet population was killed after the wreck of the *Torrey Canyon* (see p. 132). Here, *Enteromorpha* colonized exposed rocks within a year (Fig. 6.7a) and was then replaced by a canopy of *Fucus*.[235] When limpets were excluded from small areas in the Isle of Man by wire mesh fences 25 mm high, the same succession was observed during the winter and spring. In experiments that were started in July and September, however, *Enteromorpha* failed to appear before the *Fucus* canopy formed.[93] This result emphasizes that many fugitive species are delicate forms with thin thalli (see Table 1.3), and can survive in the intertidal only during the winter months.

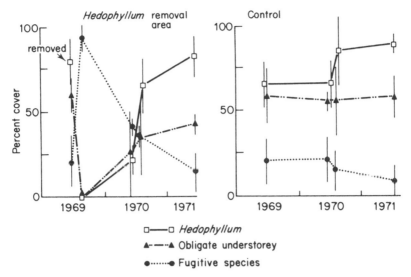

Fig. 6.8 Effects of removing the dominant kelp, *Hedophyllum sessile*, from an area of 4 m² on the coast of Washington, U.S.A.[53] (© 1978 The Ecological Society of America.)

Algal strategies: competitors, stress-tolerators and ruderals

A useful parallel can be drawn between the ecological categories of algae described by Dayton[53] and the three 'primary strategies' of terrestrial plants proposed by Grime.[85] The fugitive seaweeds are clearly equivalent to the annuals and short-lived perennials among flowering plants which are classified as **ruderals**. They have very short life-spans, rapid rates of photosynthesis and growth,[78] and most of their production is devoted either to the formation of photosynthetic tissue or to reproduction. The obligate understorey species are adapted to growth at low irradiances, and hence are shade-tolerant, in common with an important group of **stress-tolerant** terrestrial plants. Many of these algae are also found in deep sublittoral habitats beyond the lower limit of the deepest canopy-forming algae. Here, their adaptation to growth at low irradiance does not appear to include 'chromatic adaptation' (see p. 53), but probably owes more to their ability to conserve organic energy through low rates of growth and respiration and their resistance to grazing (e.g. crustose coralline algae). These characteristics are also typical of shade-tolerant flowering plants.[85]

The stress-tolerant algae also include the seaweeds of the upper shore. *Pelvetia* and *Ascophyllum*, for example, both exhibit slow growth rates, and persist in stressed conditions (*Pelvetia* — desiccation and low nutrient levels, Fig. 6.6, Table 6.3; *Ascophyllum* — intense shade, p. 130), and *Ascophyllum* also possesses the longevity and unpalatability of many stress-tolerant plants. Most of the species of *Fucus*, on the other hand, have faster growth rates (see Table 6.1) and are less tolerant of desiccation, shading or low nutrient levels. Their broader, more leaf-like thalli rapidly form an effective canopy over the rock surface, so that they easily shade out other species (see p. 130). These features are typical of the **competitors** among flowering plants, and they are displayed to an even greater extent by the laminarians. Thus, the majority of Dayton's canopy species can be classed as competitors in Grime's scheme but, strictly speaking, canopy species are **dominants**, and it is possible to distinguish **stress-tolerant dominants**, such as long-lived forest trees with shade-tolerant seedlings, as well as **competitive dominants**. *Ascophyllum* may have a short life-span in comparison with an oak tree but, nevertheless, it occupies a similar niche in the ecology of temperate rocky shores in the North Atlantic to that of oak or beech in the terrestrial ecology of the same climatic zone. Five species representing different stages in algal succession on the Pacific coast of North America have also been shown to possess different functional attributes that can be successfully related to the position of each species in the succession.[136]

Competition from animals and disturbance by wave action

The dominance of the eulittoral zone by *Ascophyllum*, like that of terrestrial communities by oak trees, is possible only in habitats that are relatively free from disturbance. For *Ascophyllum* and its fellow fucoids, this means sheltered shores in which wave action is rarely strong enough to remove whole plants, so that the turnover of individuals in the population is small. The majority of plants are, therefore, too large to be significantly affected by the herbivores in the ecosystem — none of which do more than scrape or nibble — and are large enough to exert a 'whiplash' effect on the rock surface, which prevents the settlement of barnacle larvae. If the combined effect of wave action and herbivores reduces the fucoid cover, barnacles will settle, and will then undercut the holdfasts of the algae as the calcareous plates of the adult animals spread out during growth. Thus, barnacles compete with algae for primary space on the rock surface, and they are inevitably successful in this competition, provided that other factors do not intervene. However, the barnacles are themselves out-competed for space by mussels, which always settle on a secondary substrate, such as barnacles or algae, rather than on bare rock, and then overgrow and smother whatever is underneath. Mussels appear to be the 'competitive dominant' on rocky shores in many parts of the world.[53, 238]

It is only on extremely exposed shores, however, that this succession of algae $\longrightarrow$ barnacles $\longrightarrow$ mussels normally proceeds to completion. In less exposed sites, whelks and starfish prey so heavily on the mussels that barnacles or algae are able to persist. The importance of predation for seaweed growth has been demonstrated in New England, where mussels dominate exposed headlands and the red alga *Chondrus* dominates more sheltered sites. When predators were excluded from these sheltered sites, the *Chondrus* population was replaced by mussels within 12 months (Fig. 6.9). The density of herbivorous snails also increased as wave action decreased in this locality, and this carried an additional benefit for the *Chondrus*. In more exposed sites, where snail numbers were low, fugitive

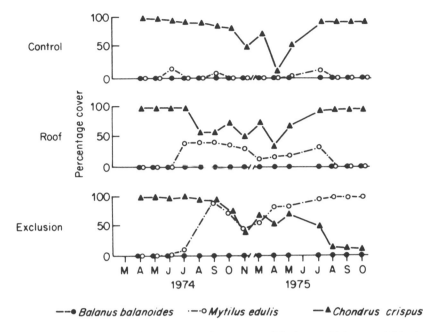

—• *Balanus balanoides*   ·—○ *Mytilus edulis*   —▲ *Chondrus crispus*

Fig. 6.9 Effect of excluding predators of the mussel (*Mytilus edulis*) from established stands of *Chondrus crispus* in the lower intertidal zone on the coast of Maine, U.S.A.[138] (© 1978 The Ecological Society of America.) 'Exclusion': wire cages fixed to rock to exclude predators; 'Roof': cages without sides as controls for the physical protection and shading caused by the exclusion treatments.

algae such as *Enteromorpha* could flourish and compete with *Chondrus*, but these species were more heavily grazed than the tougher thalli of *Chondrus*, so that decreased wave action resulted in decreased competition.[138]

Thus, severe wave action inhibits the establishment of intertidal seaweeds in two ways: by the disturbance resulting from wave action, and by the increased competition for space from sedentary animals. Few of the larger seaweeds are, therefore, to be found on exposed shores. One distinctive exception is the 'sea palm' (*Postelsia palmaeformis*) on the Pacific coast of North America, which occurs only in the midst of mussel beds in sites subjected to extreme exposure. This kelp, which looks rather like a miniature brown palm tree about 30 cm high, can apparently withstand severe wave action and compete successfully for space by growing in very dense patches (up to 300 plants m^{-2}) that smother and sometimes dislodge other algae and barnacles. Unlike most large brown algae, the individual plants are annual, but the patches persist for several years because most of the young sporophytes develop in the immediate vicinity of the parent plants.[52] These patches of *Postelsia* are, however, gradually invaded by the surrounding mussels and, in the absence of intense disturbance that will clear groups of mussels and re-start the succession, the *Postelsia* will disappear. Such intense disturbance may be caused by wave action during storms, or by additional forces such as battering of the rocks by the drift logs that are so common on these coasts. A detailed survey of one mussel bed in relation to the extent and variability of such disturbance[185] showed that *Postelsia* occurred in areas that received a moderate intensity of disturbance at frequent intervals ('disasters'), but was absent from sites with little disturbance, and from those in which large-scale disturbances occurred at infrequent and irregular intervals ('catastrophes'). This unusual seaweed — part ruderal, part competitior and part stress-tolerator — seems, therefore, to be dependent on regular disasters in the mussel population, but is unable to survive in competition with the mussels during the longer intervals between catastrophes.

These examples of the interactions between the plants, the animals and the physical environment of rocky shores illustrate the fascinating range of ecological relationships that have been revealed by recent quantitative and experimental studies in the field. More detailed accounts are given by Carefoot[27] and reported in the papers referred to in this section.

7

Geographical Distribution of Marine Plants

The basic data for studies of the geographical distribution of marine plants have been accumulating steadily — in the form of species lists for individual benthic sites and plankton samples — since the first seaweed was named. These data represent an enormous mass of information, and the first task of the phytogeographer is to ensure that the identifications and the nomenclature of different botanists, collecting throughout this time and in all parts of the world, are consistent. Only then can the overall patterns of distribution of the different species be described, and the interpretation of such patterns in terms of the ecological requirements of the plants or their evolutionary history begin. It is not surprising, perhaps, that the only world-wide analyses for whole groups of marine plants that have been completed refer to two of the smallest groups — the seagrasses [89] and the marine wood-rotting fungi. [99] This chapter describes some examples of the distribution patterns of both seaweeds and marine phytoplankton, and examines the types of problems raised by these distributions.

DISTRIBUTION OF INDIVIDUAL SPECIES OR GENERA

Patterns of distribution

Large brown algae belonging to the Laminariales (kelps) and the Fucales (fucoids) are conspicuous and frequently dominant on rocky shores throughout the temperate regions of the world (see p. 121). Most of the individual species in these two groups, however, are less widely distributed, and many are restricted to a single coastline or ocean. Among the kelps, for example, *Laminaria hyperborea* is found only in N.W. Europe, *Nereocystis luetkeana* on the Pacific coast of North America, and *Undaria pinnatifida* on the shores of China and Japan (Fig. 7.1). The giant kelp, *Macrocystis pyrifera*, is unusual among temperate seaweeds in that it is found on both sides of the equator (Fig. 7.1; see p. 147), and other species of *Macrocystis* extend the range of the genus to British Columbia and South Africa, but none of the

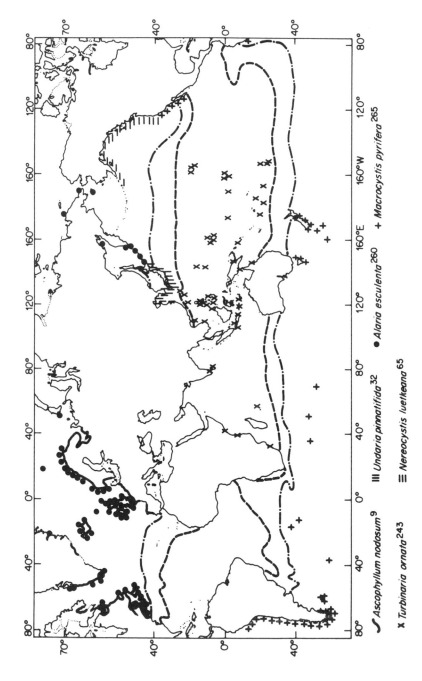

Fig. 7.1 World distribution of six brown algae of ecological and economic importance, together with the 20°C summer isotherms (— · — · —) and the 20°C winter isotherms (— — — —) for surface waters in both hemispheres.

Ascophyllum nodosum[9]   III *Undaria pinnatifida*[32]   ● *Alaria esculenta*[260]   + *Macrocystis pyrifera*[265]

X *Turbinaria ornata*[243]   ≡ *Nereocystis luetkeana*[65]

really large kelps has penetrated into the North Atlantic or the N.W. Pacific. Conversely, the numerous species of *Laminaria* are widespread in temperate seas of the northern hemisphere, but only four species with very limited distributions have been found in the southern hemisphere. [119] The genus *Fucus* is entirely confined to the northern hemisphere, and several species are restricted either to the North Atlantic, together with the common fucoid *Ascophyllum nodosum* (Fig. 7.1), or to the Pacific. Australia and New Zealand possess their own distinctive set of fucoid genera (e.g. *Hormosira*, *Acrocarpia*), many of which are not represented elsewhere in the world.

Rocky shores in the tropics present a quite different appearance from temperate shores, since all of these large, broad-bladed species of brown algae are absent. They are replaced in the intertidal by a mixed turf of smaller brown, green and red algae, and in the subtidal by the large, but more dissected fronds of the brown algae *Sargassum* and *Turbinaria*. Most of these species are not found outside the tropics (e.g. *T. ornata*, Fig. 7.1). Many of the intertidal seaweeds are also absent from Arctic and Antarctic regions simply because the rock surfaces are regularly scoured by ice, but subtidal kelps may penetrate to very high latitudes because they grow beneath the ice (e.g. *Laminaria* and *Alaria* in the Arctic; *Phyllogigas* and *Macrocystis* in the Antarctic, Fig. 7.1).

On the basis of this summary of the distribution of the larger brown seaweeds, it is possible to distinguish several patterns of distribution, which are listed with examples in Table 7.1. None of the large brown algae shows a cosmopolitan distribution, since none occurs in both tropical and cooler waters, but the genus *Ectocarpus* appears to be almost ubiquitous, along with the familiar green algae *Ulva lactuca* and *Enteromorpha compressa* and the important 'agarophyte' (i.e. agar-producing plant) *Gracilaria confervoides*. Most of the patterns of distribution that can be recognized among the

Table 7.1 Global distribution patterns of marine algae, with examples from the commoner species of seaweeds and phytoplankton.

Pattern	Phaeophyta	Rhodophyta and Chlorophyta	Phytoplankton
Cosmopolitan	*Ectocarpus* spp.	*Ulva lactuca* *Gracilaria confervoides*	*Skeletonema costatum* *Coccolithus huxleyi*
Circum-Arctic	*Alaria* spp.	–	*Thalassiosira hyalina*
North temperate	*Fucus* spp. *Laminaria* spp.	*Palmaria palmata*	–
Atlantic	*Ascophyllum*	*Chondrus crispus*	–
Pacific	*Nereocystis*	*Iridaea cordata*	*Denticula*
Tropical	*Turbinaria* spp. *Sargassum* spp.	*Eucheuma* spp. *Halimeda* spp.	*Planktoniella sol* Dinoflagellates Coccolithophorids
South temperate	*Laminaria pallida*	–	–
Australasia	*Hormosira banksii*	–	–
Circum-Antarctic	*Durvillaea antarctica*	–	–
Bipolar	*Macrocystis pyrifera*	–	*Thalassiosira antarctica*

large brown seaweeds can also be seen in familiar and economically important species of both green and red seaweeds. Thus, 'Irish moss' (*Chondrus crispus*) has a North Atlantic distribution, similar to that of *Ascophyllum*, whereas 'dulse' (*Palmaria palmata*) occurs throughout the temperate zone of the northern hemisphere. The 'carrageenophytes' *Eucheuma* (see p. 165) and *Hypnea* are tropical, while the distribution of *Iridaea* resembles that of *Nereocystis*. It is less easy to find examples of economically important red and green algae with characteristically southern distributions, since the total length of temperate shoreline in the southern hemisphere is much smaller than that in the north, and only the giant kelps are harvested commercially on an extensive scale. Nevertheless, Australasia in particular possesses a remarkably exclusive seaweed flora. Of the 1010 species recorded for Australia, 71% are not found anywhere else (i.e. they are *endemic*), and 42% of the marine algae in New Zealand are also endemic.[33]

The geographical ranges of most phytoplankton species are more difficult to define than those of seaweeds because the individual plants are unattached and thinly dispersed through the three dimensions of the oceans, instead of being concentrated into the single dimension of a shoreline. In addition, the large seasonal variations in the abundance of nearly all species of phytoplankton means that each region must be sampled very thoroughly at all times of the year before any species can be said *not* to occur there. For these reasons, it is only the more abundant phytoplankton species whose distributions can be described with any certainty. Of these, many diatoms are more or less cosmopolitan, together with *Coccolithus huxleyi* and a few dinoflagellates.[234] The majority of non-diatom species, however, are intolerant of very low temperatures, so that diatoms tend to dominate the plankton in all polar waters. Conversely, there are few diatoms, but many dinoflagellates and coccolithophorids, which are restricted to warmer waters (Table 7.1, Fig. 7.2). Thus, phytoplankton can be grouped into warm-water and cold-water species in much the same way as the benthic algae, but there are few examples of species which are restricted to a single ocean. Unlike many seaweeds, nearly all species of phytoplankton occur throughout the world in a given climatic zone, perhaps because living cells or resistant resting stages are more easily distributed by ocean currents than are the propagules of seaweeds. One notable exception to this generalization is the diatom *Denticula*, which is confined to the North Pacific. Within each ocean, however, distinct assemblages of phytoplankton species can often be recognized in oceanic, coastal and estuarine regions, and the characteristic differences between these assemblages are discussed later in this chapter (see p. 153).

Correlations between distribution and environment

The patterns of distribution described for both benthic algae and phytoplankton can be divided into those showing latitudinal boundaries (i.e. polar vs temperate vs tropical) and those with longitudinal boundaries (i.e.

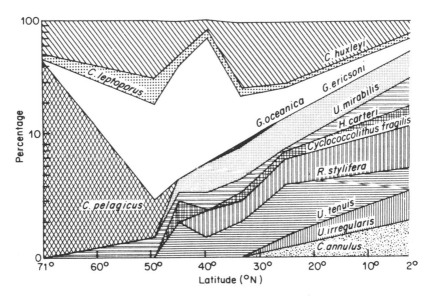

Fig. 7.2 Percentage contribution of different species of coccolithophorids to the total number of coccoliths recovered from recent sediments at various latitudes in the North Atlantic.[156] C = *Coccolithus*; G = *Gephrocapsa*; H = *Helicosphaera*; R = *Rhabdosphaera*; U = *Umbilicosphaera*.

Atlantic vs Pacific; coastal vs oceanic; etc.). The very names used to describe the patterns of the first type suggest that such distributions are related to temperature, and it has long been accepted that temperature is a primary factor controlling the geographical distribution of all groups of marine plants. Since the temperature regime in any one place is rather complex, it is necessary to identify the features of that regime which are critical for the growth, reproduction or survival of a species in order to understand the detailed relationship between temperature and distribution.

For samples of living phytoplankton, the range of temperatures at which a given species has been collected provides the best indication of the tolerance of that species towards different temperatures in the natural environment. Most species that are restricted to a particular range of latitudes show a relatively narrow temperature range, and are described as **stenothermal** (e.g. polar diatoms, −2 to 9°C; *Coccolithus pelagicus*, 7 to 14°C; *C. annulus*, 20 to 29°C; Fig. 7.2). Some polar and temperate species have been found to grow best in culture at temperatures above those at which they are normally most abundant in the sea (see p. 79). This suggests that other factors, such as nutrient supply, may interact with temperature in controlling the distribution of such species. For example, the temperature tolerance of a species may be higher under optimal nutrient conditions in culture, than at the nutrient levels that prevail in sea water. Warm-water species of phytoplankton are more difficult to grow in culture, and there is little information as yet about the reasons for their restriction to tropical waters. Cosmopolitan species are found over a wide range of temperatures (**eurythermal**, e.g. the diatom *Thalassiosira nitzschioides*, −2 to 31°C;[234] *Coccolithus huxleyi*, 2 to 29°C[156]), but this does not necessarily mean that individual cells of these species are more adaptable than those of stenothermal species. Separate strains of cosmopolitan species isolated from different geographical areas often show marked genotypic differences in physiological characteristics. Stenothermal species may be no more than similarly local strains which can be recognized as separate species because morphological differentiation has accompanied their physiological adaptation to local environmental conditions.

Unlike phytoplankton, benthic algae occupy fixed positions in the sea, and at least one stage in the life history of each species must be able to survive the least favourable season of the year. The temperature at any given instant in time is, therefore, of little relevance, and the geographical ranges of these species are usually analysed with reference to isotherms for either the warmest or the coldest month. A particularly comprehensive survey of this type[228] has shown that, out of 3350 species of red algae, 71% were restricted to regions in which the mean temperature of the warmest month did not vary by more than 5°C, and only 1 − 2% of the species could apparently tolerate a range of summer temperatures of more than 15°C. This result suggests that only a small proportion of the Rhodophyta are eurythermal and cosmopolitan in distribution. A similar approach has been applied to the distribution of coccolithophorids in the North Atlantic, and the results have been used to provide information about the changes in water temperature during the Ice Ages. The cells of these species are covered by small calcareous scales ('coccoliths') which persist in the sediments and are identifiable to the species (see p. 7). The distribution of each species in recent times (i.e. in surface sediments, Fig. 7.2) was compared with the distribution at different periods through the Pleistocene (i.e. in sediments of known radiocarbon date). The modern limit of distribution of each species was then matched with an isotherm for modern surface waters, and it was postulated that the corresponding isotherm in the Pleistocene followed the same course as the fossil distribution limit for that species. This analysis assumes that the temperature requirements of the species have not changed over the last 24 000 years, but the results based on eight different species are consistent and indicate that, during the Wisconsin glaciation, each surface water isotherm was about 15° of latitude south of its present position in the North Atlantic.[155]

This simple approach of correlating isotherms with distribution limits has the disadvantage that no single isotherm is likely to be equally relevant to all species. For example, the distribution of cold-water species will tend to be limited by high temperatures in the summer (e.g. *Macrocystis pyrifera* vs 20°C summer isotherm, Fig. 7.1), whereas tropical species will be more strongly influenced by cold temperatures in the winter (e.g. *Turbinaria ornata* vs 20°C winter isotherm, Fig. 7.1). For other species of either type, it may be the temperature tolerance of the juvenile plants, or the appropriate combination of factors for reproduction, which prevents the species from spreading any further, and not the ability of mature plants to withstand adverse temperatures. As with the vertical distribution of seaweeds on rocky shores (see p. 133), it is extremely difficult to establish the exact causes for the distribution limits of any species, and the following two examples illustrate the complexities of the problem, and the need for experiments in the field as well as in the laboratory.

Alaria esculenta has a circum-Arctic distribution (Fig. 7.1) and its southern limits in N.W. Europe are closely correlated with the 16°C summer isotherm (Fig. 7.3). This isotherm also correlates, however, with a marked change in the character of the coasts — from predominantly rocky

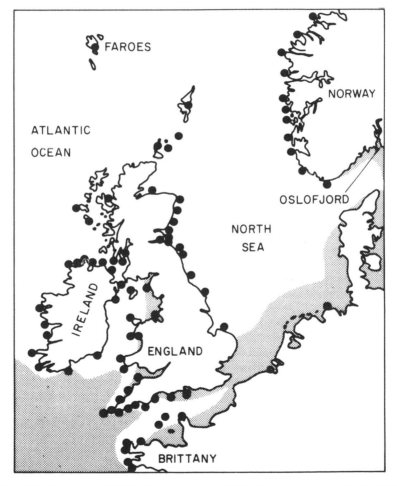

Fig. 7.3 Distribution (●) of *Alaria esculenta* in N.W. Europe, together with the 16°C isotherm for August.[242] Waters shown shaded experience surface temperatures in summer above 16°C.

shores in the north to extensive stretches of sand and shingle in the southern North Sea and in S.E. England — and divides coasts which are regularly exposed to Atlantic swell from more sheltered ones. The distribution of *Alaria* in this region could, therefore, be controlled by either of these factors, and not by temperature. To distinguish between these three possibilities, mature plants of *Alaria* were moved one autumn from the west coast of Norway to a sheltered rocky site in Oslofjord, where the species does not occur naturally. Most of the plants grew rapidly during the following spring (see Fig. 5.5) but, as the summer temperatures in the fjord rose to a maximum of 17 – 18°C, the plants stopped growing and the fronds were rapidly eroded or overgrown by other algae. By September, all of the transplanted plants had died.[242] Zoospores had been produced by the transplants in the spring, and many young sporophytes appeared in April, so that conditions in the fjord were evidently favourable for completion of the gametophyte generation. However, the young sporophytes suffered the same fate as the transplants, and were eroded, overgrown and eventually killed during the summer. Parallel culture experiments with young sporophytes showed rapid growth at 14°C, but almost no growth at 17°C, and low salinities intensified the adverse effects of high temperatures. In Oslofjord, the salinity falls to 18 – 20 ‰ at the same time as the temperature reaches 17°C, so that this factor probably contributes to the rapid degeneration and death of the plants during the summer.[242] Thus, high summer temperatures, combined with lowered salinity, can cause the death of large, established plants in a region which permits successful reproduction and establishment of juveniles, as well as the rapid growth of adult plants at other times of the year.

A different pattern of control has been demonstrated for the red alga *Bonnemaisonia hamifera* in the same region. Both gametophytes and tetrasporophytes of this heteromorphic species (see Table 5.1) are found in France and Britain, but only tetrasporophytes have been found in Scandinavia. Laboratory experiments have shown that the tetrasporophyte produces tetraspores in short-day conditions (Table 5.4) at 15°C, but not at either 10° or 20°C,[143] and it can be shown that, when the days are short enough for reproduction in Scandinavia, the temperature is too low. This means that the life history of the plant is never completed in the field, and the gametophyte generation is completely absent. The tetrasporophytes, however, can survive and reproduce vegetatively in Scandinavia, and other marine algae with complex life histories may also be represented by only one, vegetatively reproducing, phase at the extremes of their geographical range.

The problem of disjunct distributions

When the available records for a species form a solid block or a more or less continuous line on a world map, it is possible to interpret the distribution entirely in terms of the current levels of environmental factors and the physiological tolerance of the species. The distribution of some species, however, is disjunct, with two or more distinct blocks, separated by substantial geographical barriers, and such patterns cannot be explained without reference to other aspects of the history or biology of the species. *Macrocystis pyrifera*, for example, displays a **bipolar** distribution, since it occurs in the temperate zones of both hemispheres but is absent from the tropics (Fig. 7.1). The polar diatom *Thalassiosira antarctica* is also now known to be bipolar [91] although, as its specific name suggests, it was originally thought to be confined to the Antarctic. The problem is to explain how such species, which are intolerant of warm temperatures, were able to migrate through the tropics. One possibility is that, during the Ice Ages, tropical waters may have become cool enough for the growth of temperate plants. This explanation has recently been challenged on the basis of palaeotemperature measurements and it has largely been discarded by marine zoogeographers,[22] but neither of the alternative explanations of bipolarity (the 'submergence' and the 'relict' theories) seems applicable to *Macrocystis*. Photosynthetic plants are clearly unable to escape from tropical temperatures by living at greater, cooler depths because they need light, and it is inconceivable that *Macrocystis*, of all plants, could have been out-competed in the tropics by any known species of marine plant. The bipolarity of diatom species could be explained by the transport of resistant resting spores through the tropics but, if this explanation is correct, it is surprising that so few phytoplankton species are bipolar. Most of the species that are found in both hemispheres are cosmopolitan, and are also found in the tropics.[234]

Another type of disjunct distribution is displayed by numerous tropical species that are found from the Caribbean westward to East Africa, but which are absent from West Africa. *Turbinaria ornata*, for example, occurs from East Africa to the central Pacific (Fig. 7.1), and other species of *Turbinaria* reach as far as the Caribbean, but the genus is not represented at all in West Africa. The important carrageenophyte *Eucheuma* (p. 165) shows a similar pattern of distribution, along with at least six other genera of red algae, five genera of green seaweeds, several seagrasses and many marine animals.[186]

This Indo-Pacific-Caribbean pattern of distribution crosses two major geographical barriers — the wide expanse of deep sea, devoid of islands, in the eastern Pacific (the 'East Pacific Barrier'[68]), and the narrower, but equally effective, land barrier imposed by Central America. Since no shallow-water organism would be able to cross either of these barriers as they exist today, migrations must have occurred before the barriers were formed, or when they could be circumvented. Central America has presented a permanent barrier between the Caribbean and the East Pacific only since the early Pliocene (about 10 million years ago), but the East Pacific Barrier is thought to have formed during the Cretaceous (about 100 million years ago). Thus, migrations between the Indo-West-Pacific and the East Pacific regions must either have occurred before the Cretaceous or taken a route in the opposite direction through the tropical Tethys Sea, which linked the Indian Ocean with the Mediterranean and the Atlantic until much the same time as the closure of the Panama Isthmus. This latter possibility implies that some of these Indo-Pacific-Caribbean species were once completely pan-tropical in distribution, but became extinct in the eastern Atlantic and the Mediterranean as a result of the cooling of the Tethys Sea during the Miocene era.

The problem remains as to why West African coasts have not been re-colonized from either the Caribbean, or East Africa. The answer seems to lie partly in the distribution of present-day currents, and partly in historical factors. There is a strong eastward current out of the Caribbean — the Gulf Stream — but this travels well into temperate waters, where tropical organisms would perish, before its southern branch (the Canaries Current) is deflected back to N.W. Africa. Similarly, migration from East Africa around the Cape of Good Hope would be prevented by the cold Benguela Current, which flows northwards along the coast of S.W. Africa. Migration via the Red Sea and the Mediterranean seems to be more probable because the two seas are similar in temperature and salinity (see p. 21), and were joined across the Isthmus of Suez during each of the interglacial periods of the Pleistocene. However, the separation of these seas during the interven-ing glacial periods was accompanied by contrasting salinity changes: the Mediterranean became a brackish basin, and the Red Sea became a hyper-saline lake.[186] These conditions would probably have exterminated any organism that had migrated in either direction during the previous inter-glacial, and today there are very few algal species that are common to both seas — apart from those that are known to have migrated through the Suez Canal since its opening in 1869.

Some more examples of the effects of ocean currents and geological history on the distribution of marine plants are considered in the discussion of the floristic differences between phytogeographic regions (see p. 152).

The problem of overlapping distributions

Other aspects of the evolutionary history of a genus may be revealed by examining the extent to which the geographical ranges of related species overlap. Since it is considerably easier to analyse the overlap between linear ranges than between two-dimensional areas, this approach can be more easily applied to coastal than to terrestrial or planktonic organisms, and the complete seaweed flora of both coasts of the Americas has been analysed in this way.[190] The hypothesis tested was that, although related species might be expected to be concentrated in the same geographical area, they would tend to have similar ecological requirements, so that competitive exclusion would prevent substantial overlap between them. The results for 206 genera, containing an average of 6.2 species each, indicated clearly that the ranges of congeneric species overlap more frequently than would be predicted on the assumption that each species was distributed indepen-dently of the others in the genus. Within the total range for each genus, however, the ranges of the different species were distributed randomly. The latter result suggests that the congeneric species are not competing with one another and that, therefore, they must be occupying different ecological niches within the overlapping ranges. Since speciation is unlikely to have occurred within a single geographical range, it was suggested that the evolution of many of these seaweed species has involved the temporary isolation of part of the population of the parent species, followed by the re-colonization by the new species of a distinct ecological niche within the original parental range. Regular fluctuations of sea level, such as occurred during the glacial and interglacial periods of the Pleistocene, could provide a mechanism for the frequent isolation of sub-populations and the evolution of new species, which would subsequently exhibit the extensive overlap observed among congeneric species today.

This investigation provides a rare example of the quantitative analysis of seaweed distribution, and also illustrates how such information can be used to generate exciting and far-reaching hypotheses. It is true that, in this example, the hypotheses are so far-reaching that they are difficult to test, but they provide an incentive for us to obtain the genetical and ecological knowledge necessary to test them.

PHYTOGEOGRAPHICAL CLASSIFICATION OF MARINE FLORAS

An alternative approach to the biogeography of marine plants is to consider the distributions of whole floras, rather than those of individual species or genera. This approach has given rise to disagreements about how to classify the known floras into different 'types', and what to call the 'types' that result. The names used here follow the system adopted by Briggs[22] in his comprehensive survey of marine zoogeography. The world is first divided latitudinally into seven climatic zones (the polar, cold temperate, and warm temperate zones of each hemisphere, and the tropical zone), and each zone is then divided longitudinally into *biogeographic regions*, which are separated by geographical barriers. The regions may be further subdivided into a variable number of *provinces*. A marine *phytogeographic region* can best be defined as a stretch of coast, or a body of water, with a relatively homogeneous flora, which is separated by floristic dicontinuities from other areas with different homogeneous floras.[95]

An analysis of eleven seaweed floras from the N.E. Atlantic, ranging from Morocco to Spitzbergen (Fig. 7.4), illustrates what is meant by a 'floristic discontinuity'. Although there is a gradual change in the flora between Morocco and N.W. Brittany, with some southern species being replaced by more northern ones, there is no sudden alteration in the flora at any point in this range. Between N.W. Brittany and the Faroes, however, there is a sharp change — mainly a decrease in the number of species — which is regarded as a floristic discontinuity. The Clare Island flora, from the west coast of Ireland, is clearly transitional, but shows a greater affinity to the flora of N.W. Brittany than to that of the Faroes. The floras between the Faroes and Arctic Europe (i.e. the north coast of Norway and the adjacent Russian coasts) are again fairly homogeneous, but there is another sudden change

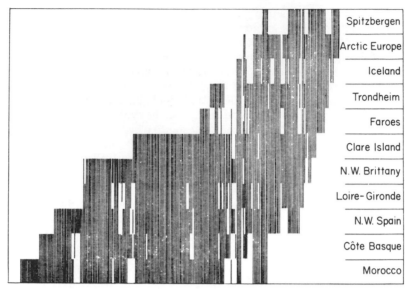

Fig. 7.4 Comparison of the benthic algal flora of 11 coasts in the N.E. Atlantic.[96] Each line represents the distribution of one species.

between Arctic Europe and Spitzbergen, which involves the disappearance of over half of the species (Fig. 7.4).

The objective definition of phytogeographic regions requires the application of quantitative techniques, similar to those discussed in connection with littoral zonations (see p. 122), to the mass of data relating to distribution. The most complete analysis of this type that is available[95] has shown that the seaweed flora of the North Atlantic can be divided into five phytogeographic regions (Fig. 7.5) on the basis of presence and absence data, such as those presented in Fig. 7.4. This analysis can be criticized on the grounds that the least significant rarity is accorded the same weight as the dominant species in the flora, and it would certainly be preferable to include abundance estimates in the analysis. However, the problems of estimating the abundance of every species on irregular rocky shores are so great that there are, as yet, too few data available for detailed analysis. Even greater problems would probably arise in the application of this approach to phytoplankton distribution, and the references to phytoplankton in the following discussion are based on less well-established generalizations about the phytoplankton flora of different regions.

Interpretation of floristic differences

The boundaries between the phytogeographic regions in the North Atlantic can be matched with winter or summer isotherms, and then interpreted in terms of the temperature tolerance of the groups of species which characterize each region (Fig. 7.5). For example, the tropical Western Atlantic

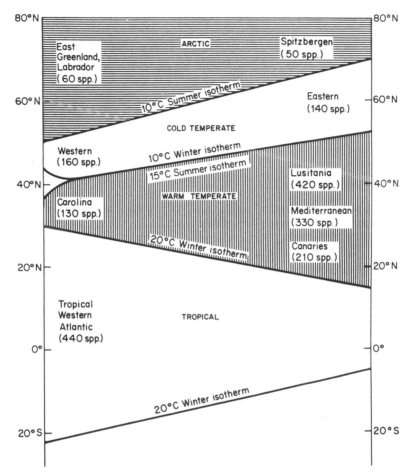

Fig. 7.5 Latitudinal range of phytogeographic regions on both sides of the North Atlantic, compared with the distribution of critical isotherms.[95]

region appears to be bounded by the 20°C winter isotherm on both sides of the equator, and this suggests that many tropical species resemble *Turbinaria* (Fig. 7.1, p. 144) in being unable to survive temperatures below 20°C. On the other hand, the boundary between the Cold Temperate (or Boreal) and the Arctic regions is marked by a summer isotherm, so that boreal species do not appear to be excluded from Arctic waters by the extreme cold of winter. The critical temperature of 10°C (Fig. 7.5) probably represents the minimum temperature at which reproduction, or the establishment of juvenile plants, can occur in many species which have their northern limits at these latitudes (e.g. *Bonnemaisonia hamifera*, see p. 146).

The Warm Temperate region occupies a wider range of latitudes in the eastern Atlantic than in the west (Fig. 7.5), and this can be largely attributed to the Gulf Stream, which carries relatively warm water towards N.W. Europe, while its southern branch (the Canaries Current) exerts a cooling effect on the coast of N.W. Africa. The greater diversity of the seaweed flora on European shores, compared with the Atlantic coast of North America (Fig. 7.5), may also be related to the dominant influence of the Gulf Stream on water movement in the temperate zone of the North Atlantic. Algal spores can be readily transported from America to Europe, but temperate species from Europe are less likely to survive the journey to America because the main westward currents flow through either tropical or polar waters.

An alternative explanation for the impoverished nature of the Warm Temperate flora of the west Atlantic, however, is that the coasts in this region are unfavourable for the growth of many seaweeds. The shores of the Carolinas are mostly sandy, and impose a substantial barrier to the migration of seaweeds along the American coast. This barrier reinforces the effects of temperature in producing a floristic discontinuity between the Warm Temperate and the Boreal regions, and a similar effect can be seen in the eastern Atlantic. Here the boundary between the same two regions coincides with the stretch of open sea between Britain and the Faroes, and the extensive sandy shores around the southern end of the North Sea. Geographical barriers may, therefore, be more important than temperature in determining the exact position of the boundaries between phytogeographic regions. Although temperature can clearly limit the distribution of individual species (see p. 143), it is unlikely that a large number of different species will have exactly the same temperature requirements. In the absence of geographical barriers, as in the open sea, temperature produces a gradual change in the flora, rather than sharp discontinuities, and phytogeographic regions are less easily defined for phytoplankton than for benthic algae.

The total number of species of either benthic algae (Figs 7.4, 7.5) or phytoplankton (e.g. Fig. 7.2) recorded at different latitudes tends to increase from the poles to the tropics. There are marked differences, however, in the representation of different taxonomic groups at different latitudes. Red algae appear to be more sensitive to high latitudes than brown algae since, in terms of number of species, the ratio of red to brown algae falls from $4-5$ in the tropics to $1-1.5$ in the Arctic.[33] Similarly, the phytoplankton flora in the tropics is usually dominated by dinoflagellates and coccolithophorids, whereas diatoms may comprise 80% of the species and 98% of the biomass in the Arctic.[207] Each of these taxonomic groups contains such a wide range of species with different ecological requirements, however, that it is difficult to extract much biological meaning from these broad generalizations, and studies of the distribution of ecological groupings of marine plants may be more informative. Among benthic algae, for example, sublittoral species tend to have narrower geographical ranges than littoral species.[190] Almost all of the floristic changes recorded at different sites in the N.E. Atlantic (Fig. 7.4) occurred among sublittoral species, and the littoral flora was more or less constant, except on the ice-scoured shores of Spitzbergen.[96] These observations suggest that the greater variability of the temperature regime in intertidal habitats (see p. 35) has resulted in the selection of species with a wider temperature tolerance than are found in the sublittoral zone.

The floristic differences that are observed in most latitudes between the phytoplankton of coastal and estuarine areas and that of the open ocean appear to be related to nutrient availability, rather than to temperature. Coastal phytoplankton usually contains a high proportion of diatoms, but dinoflagellates and coccolithophorids begin to dominate the flora as nutrient concentrations fall with increasing distance from the land.[207] This progression is similar to the seasonal succession of species that occurs in temperate waters, since the spring bloom often consists almost entirely of diatoms, but other groups become increasingly important in the more nutrient-stressed conditions of summer and autumn (see p. 82). This taxonomic analysis of the floristic changes is less useful than an interpretation based on the 'algal strategies' discussed in Chapter 6. The species that are typical of estuarine and coastal waters, which include a few dinoflagellates and the ubiquitous *Coccolithus huxleyi* as well as diatoms, tend to possess small cells and fast growth rates, and respond rapidly to changes in environmental conditions. The oceanic and late successional species, on the other hand, consist of both diatoms and dinoflagellates with larger cells and slower growth rates, together with small-celled coccolithophorids whose growth rates increase very little in response to added nutrients.[122] Many of these oceanic species possess defence mechanisms against herbivores (e.g. armour plating and poisons of dinoflagellates) or participate in symbiotic associations with nitrogen-fixing blue-green algae (e.g. the diatom *Rhizosolenia*) or with zooplankton species (e.g. diatoms and dinoflagellates).[122] All of these characteristics are typical of stress-tolerant plants, whereas the estuarine and coastal species can be classified as ruderals (see p. 135). It is less easy to identify competitors among the phytoplankton, but this may be because the appropriate combination of low stress and low disturbance never occurs in the planktonic environment.

The existing data relating to the distribution of phytoplankton are less complete than those for benthic algae, and more collection and collation of information is required to substantiate and expand the generalizations that have been discussed. Nevertheless, the recent application of the concepts of evolutionary ecology to the marine flora[122] has already generated interesting and testable hypotheses, and has emphasized that much of our physiological knowledge is based on work with fast-growing and easily cultured ruderal species. An understanding of productivity in the oceans almost certainly requires an extension of such work to the less abundant and less convenient species that grow naturally in oceanic conditions.

Life-forms of algae and marine phytogeography

Most of this chapter has been concerned with the geographical distribution of taxonomic categories of marine plants, but the reference to ecological groupings in the last section is a reminder that organisms can be classified in other ways and for other purposes (see Chapter 1). The geographical distribution of groups which have been generated by different schemes of classification will yield quite different types of information. A type of classification that attempts to integrate features of the morphology and the ecology of terrestrial plants is the scheme of 'life-forms' developed by Raunkiaer.[206] There have been several attempts to apply this concept to the algae, and one example which may be particularly helpful in the interpretation of phytogeographic patterns is the scheme summarized in Table 7.2.

Table 7.2 Algal life-forms, selected from the scheme proposed by Chapman and Chapman.[34]

Life-form	Description	Examples
Calciphykes	Species with calcite or aragonite encrustation	*Halimeda* *Lithothamnion*
Megaphykes	Attached perennials, over 5 m in length	*Macrocystis* *Nereocystis*
Mesophykes	Attached perennials, less than 5 m in length	*Fucus, Chondrus* *Polysiphonia*
Chamaephykes	Non-calcareous prostrate crusts or gelatinous colonies	*Ralfsia* *Petrocelis*
Deciduiphykes	Macroscopic attached thalli persisting for only part of year; basal portion or microscopic thallus perennial	*Scytosiphon* *Himanthalia* *Bonnemaisonia*
Therophykes	Attached thalli persisting for only part of year; rest spent as zygote, spore or unicellular generation	*Monostroma* *Acrosiphonia*
Epiphykes	External epiphytes, epizoites or parasites	*Smithora* *Harveyella*

If all of the benthic algal species in the flora for a given region are assigned a life-form according to this scheme, a 'biological spectrum' of the flora can be produced, showing the number of species with each life-form as a percentage of the total number of species. Figure 7.6 compares the results of such analyses for two temperate marine floras (New Zealand and Britain) and two tropical floras (Jamaica and Peru). It is immediately apparent that the tropical floras have a higher percentage of calciphykes and mesophykes, whereas the temperate floras contain more decidui-, thero- and epi-phykes. The prominence of deciduiphykes and therophykes in temperate floras is clearly predictable from the seasonal nature of the climate, but it is more difficult to explain the increased number of calciphykes in tropical waters and the higher representation of epiphykes in temperate waters. The latter result may simply reflect the fact that the stipes of kelps and other large brown algae provide an excellent substrate for the growth of epiphytes, and that such hospitable host plants are largely confined to temperate regions.

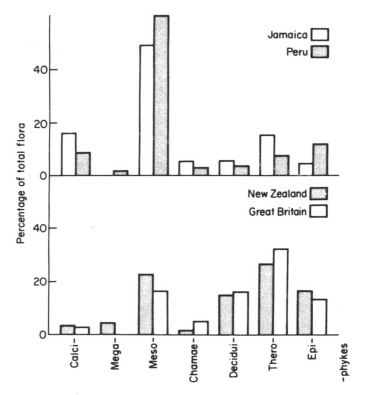

Fig. 7.6 Representation of different life-forms (see Table 7.2) in two tropical and two temperate marine algal floras.[34]

As with the definition of phytogeographic regions (see p. 150), the life-form analysis of marine floras would probably be improved by including information on the abundance and biomass of different species, rather than by treating all species as equals. Nevertheless, this type of analysis represents a fresh approach to the phytogeography of the sea, and may reveal unexpected patterns of distribution for marine algal ecologists and physiologists to investigate in detail.

8

Uses and Usage of Marine Plants

The fisheries statistics published by F.A.O. show that the total world harvest of aquatic organisms increased steadily from about 20 million tons in 1948 to about 70 million tons in 1970, but remained more or less constant from 1970 to 1977. Since three-quarters of this total harvest was marine fish, these figures suggest that the sea fisheries of the world have almost reached the limits imposed by the primary productivity of the seas (see p. 89), and there is now general agreement among fisheries scientists — if not yet among fishermen — that the total marine harvest cannot be increased substantially by hunting the familiar species of fish with ever greater intensity. Most of these commercially important fish are secondary or tertiary carnivores, however, and there would be greater scope for expansion of marine fisheries if organisms lower down the trophic pyramid could be harvested and utilized. The logical extension of this idea, of course, is to harvest marine plants rather than marine animals, but the oceanic phytoplankton that are responsible for 75% of the total marine primary productivity (see Table 4.3) are too small and too thinly distributed to be harvestable. It is, therefore, the marine macrophytes, and mainly the seaweeds, which are most likely to be of value, and this chapter is concerned with the current and potential uses for such plants, and the best ways of exploiting and managing this resource.

TRADITIONAL USES: HUMAN FOOD AND AGRICULTURE

The collection and sale of seaweeds for human consumption are now very localized in the western world, but three species of red algae are still utilized to a limited extent. These are *Porphyra umbilicalis* (laver or sloke), *Palmaria palmata* (dulse) and *Chondrus crispus* (Irish moss). In the Far East, however, the traditional use of seaweed as human food has persisted in coastal areas and has spread inland, and the cultivation and harvesting of three main products — nori, kombu and wakame — are now major industries, utilizing about one million tons of wet weed per year and with a total annual turnover equivalent to $600 million (Table 8.1). In spite of the westernization of many oriental cultures, traditional seaweed foods remain popular among racial groups of eastern origin (e.g. in Hawaii[1]), but seaweeds will probably never be an important source of human food in the west. It is thought, therefore, that the total future use of seaweed for this purpose throughout the world is unlikely to involve more than double the present amounts.[106]

There is a long history of seaweed utilization in European agriculture, mainly involving the collection of intertidal fucoids for animal fodder or for manure.[32] In some areas, such as western and northern Ireland, the growth of such plants used to be encouraged by placing large boulders in shallow, muddy bays that did not support a natural seaweed flora. These traditional practices steadily declined during the 20th century as western agriculture became increasingly industrialized, but seaweed meal is still produced (mostly from *Ascophyllum*) as an additive for cattle feed in N.W. Europe and Canada (Table 8.1). The potential for expansion of this use of seaweeds is much greater than for seaweeds as human food. An addition of 3 – 5% of seaweed meal to animal diets provides a valuable source of minerals and certain vitamins[106] and, if such additions were to be made routinely in countries with access to seaweed beds, the annual production of seaweed meal would rise to several million tons.

INDUSTRIAL USES: INORGANIC SALTS AND ORGANIC EXTRACTIVES

There have been two distinct phases in the industrial utilization of seaweeds. The first of these involved the burning of seaweeds and the extraction of sodium and potassium salts, and later iodine, from the ash. Before about 1800, seaweed ash was a major source of soda for the manufacture of

Table 8.1 Current usage of seaweeds for the food and extractives industries.[106]

PRODUCT (and algae used)	Value (10^6 \$ yr^{-1})	Weight produced (10^3 t yr^{-1})	Fresh weight harvested (10^3 t yr^{-1})
HUMAN FOOD			
Nori (*Porphyra*)	200	18	220
Wakame (*Undaria*)	200	7	60
Kombu (*Laminaria*)	200	100	700
ANIMAL FODDER			
Seaweed meal (*Ascophyllum*)	10	30	100
SEAWEED EXTRACTIVES			
Agar (*Gelidium, Gracilaria*)	40	6	150
Carrageenan (*Chondrus, Gigartina, Eucheuma*)	60	10	130
Alginates (*Macrocystis, Laminaria, Ascophyllum*)	65	15	400

soaps and glass, but it was rapidly made obsolete for this purpose by the development of a chemical process for sodium carbonate at the beginning of the 19th century. The kelp industry continued to produce potash and iodine, however, until the 1930s, and the first large-scale harvesting of the giant kelp *Macrocystis* on the Pacific coast of North America was started for this purpose as late as 1910.[32] Nevertheless, the costs of harvesting and drying the weed gradually made the industry uncompetitive as a source of these relatively low-value products, and nowadays the industrial use of seaweeds is based entirely on different products of higher value. These are a series of organic compounds which are found only in seaweed thalli.

Chemistry of seaweed extractives

The seaweed extractives of commercial importance fall into three main groups, two of which (agar and carrageenans) are derived from red algae, and the third (alginates) from brown algae. The most important genera harvested for these extractives are shown in Table 8.1, together with the total amount and value of the products. All three types of extractive are associated with the cell walls of the algae and resemble cellulose in basic molecular organization. Unlike cellulose, however, the long-chain polysaccharide molecules are composed of a variable mixture of different sugar residues. Extracts from different species, or from different parts of a single plant, or from the same species at different times of the year, show slight differences in composition, which often have significant effects on the properties and value of the product.

Agar can be divided into two components, agarose and agaropectin. The former is relatively simple and constant in composition, consisting of alternating units of D-galactose and 3,6-anhydro-L-galactose (Fig. 8.1a), none of which are sulphated. Agaropectin has the same basic structure, but with a variable degree of substitution by sulphate, pyruvate and other groups.[21] It is the ratio between agarose and agaropectin that varies from species to species, and the gel-strength of the agar is proportional to its agarose content.

Carrageenans are also composed of alternating units of modified galactose, but all of the sugar units are in the D-form, and at least one unit in each pair is usually sulphated. Several different types of carrageenan have been isolated, but there are two basic groups, designated by the Greek letters $\varkappa$-(kappa) and λ- (lambda). In the gelling constituents of carrageenan (the $\varkappa$-carrageenans), alternate units of galactose are usually present as 3,6-anhydro-galactose (Fig. 8.1b), as in agarose, but this sugar anhydride is absent from the non-gelling λ-carrageenans (Fig. 8.1c). The $\varkappa$-carrageenans are thought to form gels because the macromolecules occur in pairs as double helices (Fig. 8.1d), whereas the λ-carrageenans consist of flat unpaired chains as a result of the substitution of galactose for anhydro-galactose.[208] An occasional substitution of the same type occurs in $\varkappa$-carrageenans and causes a 'kink' in the molecule, so that the chain bends

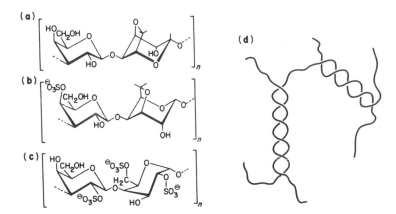

Fig. 8.1 Chemical structure of extractives from red algae. (a) Agarose (1,3 linked β-D-galactose and 1,4 linked 3,6-anhydro-α-L-galactose); (b) ϰ-carrageenan (1,3 linked β-D-galactose 4-sulphate and 1,4 linked 3,6-anhydro-α-D-galactose); (c) λ-carrageenan (1,3 linked β-D-galactose 2-sulphate and 1,4 linked α-D-galactose 2,6-disulphate); (d) proposed arrangement of polysaccharide chains in agarose and ϰ-carrageenan gels.[208]

away from its original partner and becomes paired with another molecule, and this results in a 3-dimensional network of molecules rather than a collection of independent double helices (Fig. 8.1d).

Alginates are salts of alginic acid, and the long unbranched chains contain two types of acidic sugar residues, D-mannuronic acid (Fig. 8.2a) and L-guluronic acid (Fig. 8.2b). These two components do not alternate regularly in the molecules, but occur in blocks consisting of about 20 units of one of the acids on its own, linked by regions containing both acids. The proportion of mannuronic acid is higher in young cell walls and in flexible parts of the plant, such as the blade of laminarians, whereas the guluronic acid content increases in mature cell walls and in the stipe and holdfast.[189] Blocks of guluronic acid accounted for 50% of the total alginate in attached holdfasts of *Laminaria digitata*, but only 35% in young unattached haptera.[240] Alginate gels are not formed by cooling, as with agar and carrageenans, but by the addition of calcium ions, and the greater gel strength of alginates rich in guluronic acid is related to this effect. The buckling of the molecular chain is more pronounced in blocks of guluronic acid than in those of mannuronic acid (see Fig. 8.2a, b), and this buckling provides room for the calcium ions which then link adjacent chains in what has been described as an 'egg-box' structure (Fig. 8.2c).[189, 208]

The production of seaweed extractives and their uses

Before about 1930, agar was the only seaweed extractive produced on a substantial scale, and its production was confined to Japan, where it took the form of a cottage industry, conducted by farmers during the winter. The seasonal nature of the industry was dictated by the purification procedure,

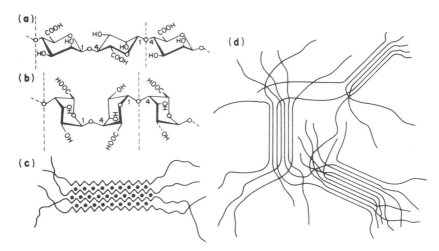

Fig. 8.2   Chemical structure of alginates. **(a)** Chain of 1,4 linked β - D-mannuronic acid residues; **(b)** chain of 1,4 linked α-L-guluronic acid residues; **(c)** association between Ca^{++} ions (●) and blocks of guluronic acid in alginate gels; **(d)** proposed structure of an alginate gel.[148, 208]

which involved — and still involves — freezing the crude extract obtained by boiling the seaweed in water. As the frozen extract thaws, the water separates from the agar and carries with it soluble impurities, such as inorganic salts and phycobilin pigments. The primary extraction of carrageenan also involves heating the appropriate seaweeds in water, but the product is precipitated from the crude extract by the addition of propanol. Alginate extraction is a more elaborate chemical procedure. The washed and macerated brown algae are digested with sodium carbonate, and the filtrate is added slowly to a concentrated solution of calcium chloride. A fibrous precipitate of calcium alginate is formed, which is converted to alginic acid by treatment with dilute hydrochloric acid. After further purification and chemical treatment, the product is marketed, usually in a powdered form as sodium, potassium or ammonium alginate. Further details of these extraction procedures are given by Chapman[32] and by Booth.[21]

These two authors also provide a comprehensive outline of the wide range of products in which agar, carrageenans and alginates are used, and it is possible here only to indicate the main properties of these substances that are of commercial value. All three types of extractive produce firm gels at low concentrations, and they are used for this purpose in numerous processed foods, confectionery, cosmetics and pharmaceutical products. Thin surface films are also used to improve the texture and quality of paper, textiles and leather goods, and sheets of calcium alginate have been coloured and flavoured and offered to the public — without much success — as substitute, cancer-free tobacco. Seaweed extractives also assist in the formation and stabilization of colloidal suspensions ranging from ice-creams and dried-milk products to preparations of drugs and emulsion paints. Artificial fibres can also be made from alginates, although their use is somewhat limited at present because they are readily dissolved by the alkalis in soap. Modern lubricants and adhesives often contain alginates and carrageenans, and another important and traditional use is in 'fining' or clarifying beers and wines.

The applications for these products have increased enormously in the last 20 years but, apart from the familiar use of agar as a microbiological culture medium, few of these applications can be said to have made a major contribution to man or his civilization. Seaweed extractives may provide an attractive artifical froth for beer and confectionery, and help convenience foods to become even more convenient, but quite different applications of seaweeds may soon completely overshadow the extractives, both in social and economic importance.

POTENTIAL USES: WASTE TREATMENT AND BIOMASS ENERGY SOURCE

Among the major problems faced by human civilization are the development of renewable sources of energy and environmentally acceptable treatments for domestic and agricultural wastes. Seaweeds can contribute to the solution of both of these problems. The potential role of microscopic algae in sewage treatment — to remove the inorganic nutrients released by bacterial oxidation of organic compounds, and to supply the oxygen for this oxidation — has long been recognized, but there is as yet no cheap and efficient means of harvesting such algae on a commercial scale. Consequently, macroscopic algae, which are much easier to harvest, are now being considered for this purpose. Several red algae with high agar or carrageenan contents have been grown successfully on sewage effluent diluted with 3 – 5 volumes of sea water,[220] but it is unlikely that the extractives industry will ever be able to utilize all of the potential production from waste treatment plants of this type. An alternative use for the seaweed biomass is to convert it to methane by anaerobic bacterial fermentation, thus providing a renewable replacement for the dwindling supplies of natural gas from fossil sources. The potential demand for such 'biomass energy' is far greater than could be satisfied by waste treatment alone, and so the collection or cultivation of marine algae specifically for their energy content is also envisaged. This could involve simple harvesting of natural populations of seaweeds, but active cultivation of marine algae in the sea ('mariculture') would probably be required.

Harvesting and cultivation represent two distinct approaches to seaweed usage, and a few examples of each, ranging from the traditional to the futuristic, are described in the rest of this chapter.

HARVESTING NATURAL PLANT POPULATIONS

Hand-harvesting

The majority of seaweed species are harvested by hand, simply because of the difficulty of designing a mechanical harvester that can operate under-water and over the irregular terrain that characterizes most seaweed habitats. The object of this section is to examine the impact of such harvesting on natural populations, and the ways in which the long-term yield of seaweed beds may be maximized. *Ascophyllum* is one of the most widely harvested species in the North Atlantic, but its longevity (see p. 130) makes it peculiarly susceptible to over-harvesting. Complete removal of a previously unexploited population is equivalent to cutting down a forest of 20-year-old trees, and recovery may take even longer than this analogy indicates because *Ascophyllum* is a poor competitor against *Fucus* in the early stages of re-colonization (see p. 130). The harvesting practice that is recommended, therefore, is to leave a stump of at least 10 cm (representing about 2% of the standing crop), so that regeneration of the original plants can occur and the population can be harvested again after 3–4 years.[227] A theoretical model of an *Ascophyllum* population subjected to different intensities and frequencies of harvesting, however, has suggested that the greatest long-term yield would be obtained by leaving over 15% of the standing crop uncut, in order to obtain even more rapid recovery, and harvesting every 2 years[227] (Fig. 8.3). The more frequent harvests are an essential part of this strategy because self-shading will occur when population densities reach a critical value, and the growth rate will decrease. This modelling approach is similar to that used for many years in the management of fish stocks,[49] and could become increasingly important in the effective harvesting of seaweeds.

Another aspect of seaweed biology that may have important implications for harvesting procedures is the seasonal variation that often occurs in the percentage dry matter or the chemical composition of the thalli (e.g. *Laminaria*, see pp. 79, 85). The content and quality of all types of extractive tend to vary with the season and the age of the plant,[32] so that maximizing the fresh weight of the harvest does not necessarily maximize the financial return.

Mechanical harvesting

The simplest type of mechanical harvester consists of a heavy rake or dredge, usually with a basket or net attached, which is dragged along the sea bottom from a boat. These rakes are currently used for harvesting lower littoral and upper sublittoral populations of *Chondrus* in Nova Scotia (Fig. 8.4), but they have been shown to remove a higher percentage of juvenile plants and plants with holdfasts than the traditional hand-harvesting technique, which uses a simple, long-handled rake.[197] As a result, regeneration occurs more slowly, and this has probably contributed to the substantially lower standing crops in mechanically harvested areas. The dredges are now being re-designed so that they cause less damage to the harvested populations. *Laminaria* is harvested in Norway using similar dredges, which cut the stipe 5–20 cm above the holdfast. No regeneration of the harvested plants is possible, but the dredge leaves the young sporophytes

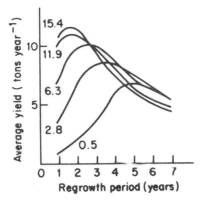

Fig. 8.3 Mathematical model of a harvested population of *Ascophyllum*: average annual long-term yield as a function of the length of time between harvests ('regrowth period') when different percentages of the standing crop are left uncut at each harvest.[227]

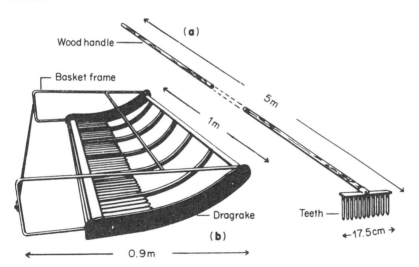

Fig. 8.4 Types of rake employed in the harvesting of *Chondrus crispus* in eastern Canada. (a) Hand-rake; (b) drag-rake towed behind a boat — the frame is covered with a 2.5 cm mesh screen.[197]

untouched, and these restore the original plant density in 3–4 years.

These dredging techniques inevitably require substantial amounts of fuel energy simply to drag the harvester along an irregular sea-bed and, if the harvested weed is to be used for energy production, these techniques are unlikely to prove economically viable. A more efficient harvester for *Laminaria* was designed and built by the Scottish Institute for Seaweed Research, which resembled a conventional dredger with a continuous belt of hooks and wire mesh replacing the chain of buckets.[101] This belt tore the seaweed from the sea-bed and carried it up into the boat, but the irregularities of the sea bottom caused substantial mechanical problems, and the development was not continued. The only really successful mechanical harvesters are those designed for the giant kelp *Macrocystis*. This species is very much easier to harvest because it is found in water depths of 10 – 15 m, and the long fronds grow up to and along the surface of the sea. The harvesters are large barges (see Fig. 8.7, p. 168) with a mowing bar that spans the full width of the ship (6 – 7 m) and cuts the fronds about 1.5 m below the water surface. A conveyor belt system carries the cut weed up out of the water and deposits it in the well of the ship. The harvesting operation has been compared with mowing a lawn (if one can imagine a lawn that is 10 m high and cut from the air!) and the effect on the plants seems to be similar in that harvesting reduces the self-shading within the population, and regeneration occurs rapidly from the cut fronds and from uncut juvenile plants. A single kelp bed can usually be cut every 4 – 6 months.

This analogy between a *Macrocystis* bed and a lawn suggests that regularly harvested seaweed populations are not strictly 'natural populations', but are 'managed populations'. The effects of harvesting on natural marine communities are likely, therefore, to be similar to the ecological impact of many agricultural practices on terrestrial communities: a decrease in overall species diversity, and a shift from stress-tolerant towards ruderal and competitive species (see p. 135). These trends have been demonstrated after a single experimental harvesting of *Ascophyllum* in Northern Ireland.[17] Thus, although the removal of a canopy species may have little direct effect on energy flow in a littoral ecosystem, because so few animals graze the larger seaweeds, harvesting cannot be regarded as ecologically harmless. As the demand for seaweeds for various uses becomes greater, it is important to designate substantial areas of rocky shore as marine nature reserves, where the diversity and the large gene pool of benthic marine organisms can be maintained.

CULTIVATION OF MARINE PLANTS

A major limitation on the maximum harvest of marine plants from natural or managed populations is the total area of firm substrate that is available for the growth of such plants. The objective of many schemes for cultivating marine plants, therefore, is simply to increase the area of suitable substrate in coastal waters. The alternative approach of tank culture, which involves growing plants in more controlled conditions on shore, requires less manpower, but demands more energy and capital investment.

Cultivation in the sea

The simplest technique for encouraging seaweed growth is to place large stones or boulders on a sandy or muddy sea bottom, and this used to be practised in some parts of the western world (e.g. Ireland, see p. 157) as well as being a traditional method for cultivating *Laminaria* and agar-producing plants (e.g. *Gelidium, Gracilaria*) in Japan. An improvement on this technique is to use an artificial substrate that can be moved, since this makes harvesting very much easier and permits a greater control of the density and composition of the algal crop. Japanese fishermen have planted bamboo or brushwood 'hibi' in shallow waters since the 17th century to increase the substrate for *Porphyra* ('nori'), and the 'hibi' were usually placed in rocky areas at the beginning of the autumn, where they collected a good crop of spores, and were then moved to sandy bays for the growth of the plants during the winter. However, the thin leafy thalli disappeared in the summer and, until the life history of *Porphyra* was established in 1949, it was impossible to predict or control the size of the crop that reappeared in the autumn. The discovery by a British phycologist[50] that the filamentous, shell-boring alga *Conchocelis* was an alternate phase in the life history of *Porphyra* (see Table 5.1) had an immediate and dramatic impact on the size and the reliability of the cultivated crop in Japan. The conchocelis-phase could now be cultivated in tanks during the summer, and the results of more recent physiological work on the effects of temperature and daylength (see p. 111) have permitted precise control of spore production by the conchocelis. The brushwood 'hibi' have been replaced by rope or nylon nets, which are soaked in the spore-laden water from the culture tanks and then spread out over 67 000 ha of Japanese coastal waters.[10, 258]

Academic studies of algal life histories (see Table 5.1) have also had a major impact on the cultivation of the two genera of kelps that are important foodstuffs in China and Japan. *Laminaria* and *Undaria* ('kombu' and 'wakame', Table 8.1) are now extensively grown by a 'forced cultivation' technique that involves controlling the reproduction of the microscopic gametophyte generation through irradiance and light quality (see p. 112), and allowing the young sporophytes to grow on in culture until they are large enough to be transferred to the sea. There, they are grown on vertical ropes hanging from bamboo rafts (Fig. 8.5) and plants that take two years to mature in natural conditions can be raised to harvestable size in one year.[90] The growth of the gametophytes and young sporophytes in culture also means that these species can be cultivated in areas where the summer temperatures are too high for the development of natural populations (see p. 146), and artificial selection has recently produced a strain of *Laminaria* that is less susceptible to high temperatures, so that a further spread of cultivation into warm waters may be possible.[36]

The gradual improvement of natural stocks by artificial selection is also an important aspect of a recent attempt to establish seaweed cultivation in the Philippines. The cultivation techniques being used here for the carrageenophyte *Eucheuma* differ from those for *Porphyra* and *Laminaria* since they rely on vegetative propagation. Large plants are divided into small pieces, which are then tied individually onto horizontal nets in the upper subtidal zone.[56] The selection procedure is simple. The smaller plants are sold at each harvest, whereas the larger — and therefore fast-growing — plants are divided up to provide the 'seed' for the next crop. Since there is no tradition of seaweed cultivation in the Philippines, the local phycologists

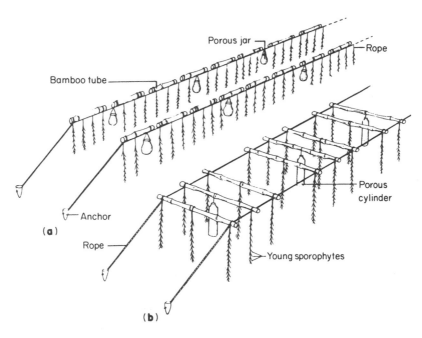

Fig. 8.5 Cultivation of *Laminaria* in China: single- and double-line bamboo rafts, fitted with porous earthenware jars or cylinders for slow fertilizer seepage.[36]

face the interesting challenge of explaining the elements of seaweed physiology and ecology in pictures (Fig. 8.6) to help the local population to find the best new sites for farming *Eucheuma*.

The intensive cultivation of seaweeds by these methods is beset by many of the problems associated with crop growth on land, such as nutrient deficiencies and the control of diseases. Cultivated *Porphyra*, for example, is subject to a 'red rot' caused by a marine species of the fungus *Pythium*.[4] Pouring inorganic fertilizers into the sea will clearly do little to relieve the nutrient problem, but the use of earthenware pots (Fig. 8.5) or solid pellets containing fertilizer, which slowly release nitrates and phosphates into the water, has proved effective in the cultivation of *Laminaria*[36] and *Gelidium*.[267] An alternative technique for young plants is to immerse them in a nutrient solution for a short period and then return them to the sea. The plants absorb enough nutrients in 30 minutes to last them for 7 – 10 days.[258] The traditional areas for *Porphyra* cultivation in Japan are mainly shallow estuaries which receive substantial nutrients from domestic sewage, but increased industrialization has closed some of these areas because of pollution by heavy metals.

Tank culture and 'high technology' in seaweed cultivation

The development of seaweed cultivation in North America has occurred

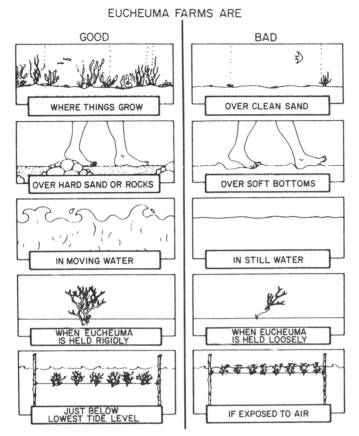

Fig. 8.6 Cultivation of *Eucheuma* in the Philippines: poster produced as an aid in teaching the local fishermen how to select the best sites for new *Eucheuma* farms.[56]

largely in response to the demand created by the agar and carrageenan industries, and has, therefore, concentrated on red algae such as *Chondrus* and *Gracilaria*. These are usually grown as unattached plants, which are kept in suspension by vigorous agitation of the water in the culture tanks. The plants are propagated vegetatively, so that suitable strains for these rather unnatural growth conditions can be rapidly selected by cloning individual plants. Selection is primarily for rapid growth, but resistance to epiphyte growth and infrequent sporulation are also desirable characteristics in cultivated strains. Temperature control is obtained in some systems by placing the culture tanks in glasshouses,[177] and the high nutrient demand of the rapidly growing plants may be met by enriching the sea water with sewage effluent (see p. 161). This serves the dual purpose of lowering the costs of cultivation, and providing an effective waste treatment system. Most of these projects are still at the pilot plant stage, but the average yields obtained over periods of up to 7 months are at least as high as some of the best agricultural yields.[220] Whether these yields can be maintained when the systems are scaled up to full-sized production plants, and whether the costs can be reduced sufficiently to make the projects economic, have yet to be established.

The demand for alginates in the foreseeable future is not great enough to require the cultivation of brown algae in western countries, but the prospect of using seaweeds as a renewable energy resource (see p. 161) has inspired an imaginative project for cultivating *Macrocystis* in the open ocean.[261] An open mesh of stout plastic lines 15–30 m below the surface would provide the artificial substrate for *Macrocystis* plants, and a wave-activated pump would raise nutrient-rich water from a depth of about 300 m to the surface in order to support the rapid growth of the kelps (Fig. 8.7). Field experiments in deep waters off the Californian coast have already shown that rapid growth of *Macrocystis* can be maintained in a mixture of surface water and artificially upwelled water from below the thermocline.[179] Even so, it would require an ocean farm 160 km in diameter to supply just 2% of the total U.S. consumption of energy in 1970[264] at a cost which is unlikely to be competitive with other sources of energy for some time to come.

This Ocean Farm Project seems at the moment to be in the realms of seaweed science fiction, and most of the agitation culture systems could not be used to supply biomass for energy, since more energy is required to run the

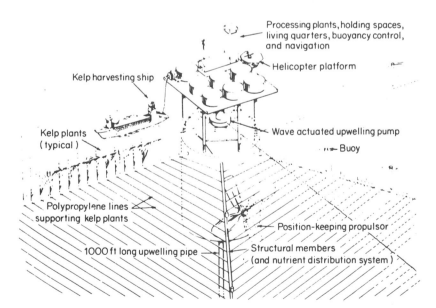

Fig. 8.7 Drawing of a project for an ocean farm in which *Macrocystis* would be grown in artificially upwelled deep ocean water.[261] Economically viable farms would need to cover an area of at least 8000 ha, and would be held in position by the propulsion unit, or be allowed to drift with natural ocean currents.

plants than can be obtained from the algae that are grown in them. The current emphasis, therefore, is on more modest and cost-effective projects, but it is clear that seaweeds have become part of the technological revolution, and that applied marine botany will soon be a subject to be reckoned with.

9

Bacteria and Fungi in the Sea

The bacteria and fungi that live in the sea cover the full range of morphology, life history and physiology that is found in these organisms. All four of the major classes of fungi are represented in the sea (see Table 1.1), and marine bacteria include photosynthetic and chemosynthetic species, as well as aerobic and anaerobic heterotrophs. Thus, there is little, apart from their ecology, to distinguish marine microorganisms from those of terrestrial and freshwater habitats. This contrasts with the situation among flowering plants (see Chapter 1) and suggests that life in the sea is not fundamentally different from life on land for bacteria and fungi, and that the transition from sea to land — and back again — may have occurred many times during the evolution of these groups.[125] Only about 1% of the known species of fungi are found in the sea, but this is not thought to mean that the marine environment is physically unfavourable for fungal growth. It is usually attributed to the restricted amount and variety of organic substrates available in the sea, and to the low selective pressure exerted by the relatively constant environment.[125] The same arguments probably apply to the bacteria but the greater difficulty of defining a species in this group makes it impossible to obtain a reliable estimate of the numbers of bacterial species in either marine or terrestrial habitats.

Since, therefore, the basic distinction between marine and terrestrial microorganisms lies in their ecology, it is the contribution of these organisms to the functioning of marine ecosystems, and the ways in which they interact with other marine plants, that are emphasized in this chapter. Other aspects of marine microbial biology are described in a comprehensive and well-illustrated book by Sieburth.[232]

MICROBIAL ACTIVITIES IN MARINE ECOSYSTEMS

The majority of bacteria and fungi are heterotrophic organisms whose primary role in all ecosystems is generally considered to be the decomposition of organic detritus and the release of inorganic nutrients. Although marine microorganisms certainly assist in the decomposition of organic matter, it seems unlikely that, under natural conditions, they release many inorganic nutrients in the process. The N:C and P:C ratios in most bacteria are higher than in much of the plant material that they decompose, so that they must accumulate, rather than release, nitrogen and phosphorus, and they may even compete with the autotrophs for additional supplies of these elements in the water.[192] The recycling of nutrients in aquatic ecosystems need not involve complete remineralization, because many of the autotrophs can utilize organic sources of nutrients, such as typical excretory products of planktonic animals (see p. 69). Therefore, the most important contribution of bacteria to the marine ecosystem is probably the conversion of organic wastes into forms that can re-enter the food chain. For example, much marine detritus, such as fragments of seaweed thalli, consists mainly of carbohydrate and has only limited nutritive value. The growth of bacteria and fungi enriches such particles with the protein and mineral content of the microbial protoplasm, and makes them more attractive for animal consumption. In addition, a large proportion of the primary production in the sea may be released as dissolved organic compounds (e.g. extracellular products of phytoplankton, p. 77; mucilages of many seaweeds), which cannot be used directly by animals. However, the high surface:volume ratio of bacteria enables them to assimilate this dissolved organic carbon (DOC), even at the low concentrations found in the sea, and the bacterial growth provides a suitable food source for a wide range of consumers.

Thus, the primary result of bacterial and fungal activity in aerobic marine habitats is the stimulation of energy flow and, in inshore waters receiving a large input of seaweed detritus, bacterial production may exceed that of the phytoplankton. Nutrient regeneration is a secondary and indirect consequence of microbial activity in such ecosystems. In anoxic habitats, such as undisturbed sediments, however, anaerobic heterotrophs and autotrophic bacteria exert more direct effects on the concentrations of certain nutrients (see below), and these effects may extend to the water above.

MARINE HABITATS OCCUPIED BY BACTERIA AND FUNGI

Planktonic and neustonic habitats

The decomposers in sea water can be isolated and identified by incubating water samples in favourable laboratory conditions, and studying the bacterial or fungal colonies that develop from individual cells or spores in the water. This technique fails, however, to distinguish between cells that were active in the plankton at the time of collection, and those that were dormant because they normally occur attached to surfaces or in sediments, or even in freshwater or terrestrial habitats. It is important, therefore, to measure microbial activity *in situ*, and this is best done by exposing natural planktonic communities to low concentrations of radioactively labelled, dissolved organic compounds. Bacteria are the only organisms to become significantly labelled under these conditions, and their numbers can be estimated by autoradiography.[98] The photic zone frequently contains bacterial populations of $10^5 - 10^6$ cells ml^{-1}, which absorb and utilize the extracellular products of the phytoplankton. Below the photic zone, the concentration of DOC decreases to about one-third of that nearer the surface, and much of the material in deep waters appears to be relatively indigestible, and may be several centuries old.[232] The physical conditions in deep water are also less favourable for biological activity (see below) and bacterial populations at depth are about two orders of magnitude smaller than those in the photic zone (Table 9.1).

Conversely, the concentration of DOC and the density of bacteria increase markedly in an ultra-thin layer at the surface of the sea. Numerous organic compounds, such as hydrocarbons, lipids and polysaccharide-protein complexes, have surfactant properties and form monomolecular layers at the boundary between air and water. Air bubbles that are forced into the water column by wave action become coated with such compounds, and these are carried to the surface as the bubbles rise. The microlayer at the surface is no more than 0.1 μm in thickness, but the concentration of DOC may be 1000 times that in subsurface water.[232] Bacterial numbers are correspondingly high — up to 10^8 cells ml^{-1} — and their productivity supports a variety of specialized, surface-living animals, which are collectively known as the **neuston**. This habitat is occasionally dominated by the hydrocarbons from a spillage of crude oil, and the yeasts and filamentous fungi that degrade hydrocarbons are sometimes found at high densities in coastal waters. However, the rate of degradation is slow in the sea, possibly because there are so few nutrients available, and marine fungi appear to contribute little to the recovery of the sea from oil pollution.[125]

Bacterial growth in the plankton and the neuston is largely dependent on extracellular photosynthate released by phytoplankton cells, but the autotrophs may also benefit from organic compounds which are synthesized and released by the bacteria. The most important of these is probably vitamin B_{12}, which is required by many species of phytoplankton, as well as some benthic algae (see p. 69). However, bacteria also synthesize compounds which are chelators of various metal ions, and may exert a significant influence on the mineral nutrition of phytoplankton through such substances (see p. 68).

Table 9.1 Densities of planktonic bacteria in water samples from five depths in the north central Pacific Ocean, and growth rates of isolated cells at natural temperatures and pressures in unsupplemented sea water.[28]

Depth (m)	Water temperature (°C)	Hydrostatic pressure (atm)	Bacterial density (cells ml^{-1})	Growth rate (doubling time, h)
1	22.6	1	1.4×10^5	14
75	18.8	7.5	5.1×10^4	11
500	7.5	50	2.0×10^4	67
1500	2.5	150	7.2×10^3	145
5550	1.5	555	5.0×10^3	210

Epibiotic and endobiotic habitats

Almost any surface that is placed in the sea will be rapidly covered — often within a day — by a film of bacteria, and this represents the first phase in the process of 'marine fouling'. Marine organisms themselves are not immune from this process, and seaweeds may carry up to $10^5 - 10^6$ bacteria cm^{-2}, often as a 'lawn' of tightly packed, upright rods. Some seaweeds produce antibiotic substances, however, which inhibit the settlement and growth of bacteria, and the growing tips of fucoids, such as *Sargassum* and *Ascophyllum*, are often completely free of microorganisms.[48] The younger, basal portions of kelp fronds also carry low bacterial populations during periods of active growth (see p. 83), although the numbers increase in the summer and autumn (Fig. 9.1). Bacterial densities tend to increase on older tissues, and the eroding tips of the fronds of *Laminaria* carry dense populations throughout the year (Fig. 9.1). Thus, the bacterial enrichment of kelp fragments begins before the thallus is broken up, and a similar process is evident in the production of detritus from the leaves of seagrasses.[88]

The presence of a bacterial 'lawn' appears to be essential for the establishment of many of the larger fouling organisms — benthic algae and sedentary animals — since regular removal of the bacterial film by brushing inhibits the fouling process completely.[232] It is possible that these effects are due to organic compounds released by the bacteria, which influence the morphogenesis of settling algae (e.g. *Ulva*, see p. 106) or animals (e.g. the coelenterate *Cassiopeia*[178]).

The filamentous morphology of many fungi enables them to penetrate organic substrates more easily than bacteria, and fungi are particularly important in endobiotic habitats. The most widely available (and most widely studied) substrate for marine fungi is wood, and over half of the known species of higher marine filamentous fungi have been collected on

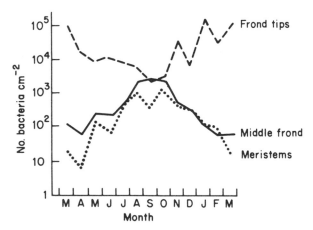

Fig. 9.1 Seasonal variations in the numbers of bacteria on different regions of the fronds of *Laminaria longicruris* (redrawn from Laycock[131]).

pilings, wrecks, experimentally-submerged panels or natural driftwood.[125] Few phycomycetes are found on wood in the sea, but these lower fungi are frequently the agents of disease in phytoplankton and in marine animals. Fish, molluscs and crustaceans are all subject to fungal diseases, which may reach epidemic proportions in the crowded, artifical conditions of aquaculture.[2] Similar diseases also occur in cultivated seaweeds. (e.g. *Pythium* on *Porphyra*, see p. 166), and numerous parasitic Ascomycetes have been isolated from natural populations of benthic algae. The symptoms of fungal attack may be gall-formation or discoloration of the tissues, but other 'weaker' parasites have no effect on the external appearance of the host, except when inconspicuous fruiting bodies are formed. Some species of weak parasites have been found in every individual of the host species that has been examined (e.g. *Mycosphaerella ascophylli* in *Ascophyllum*), and this has led to the suggestion that they represent a lichen-like association, since neither species appears to live without the other.[125] However, the absolute dependence of *Ascophyllum* on its fungal 'partner' has yet to be clearly demonstrated, and the validity of this peculiarly marine intermediate between mutualism and parasitism requires more detailed investigation.

Benthic habitats

All benthic habitats receive a more or less continuous rain of organic and inorganic detritus from the waters above them and, in sites which are not scoured by tides or deep-water currents, this detritus will accumulate as sediments. The organic content of these sediments will depend on the productivity of the water column and its depth. The primary production of the photic zone is utilized by consumers and decomposers as it sinks through the water, and relatively little reaches the sea floor of deep ocean basins. The sediments in shallower coastal waters and estuaries, however, receive a high organic input, and this supports a richer and more varied microflora than is found in the deep sea.

The depth profiles of certain chemicals in undisturbed sediments (Fig. 9.2) reflect the succession of microbiological processes that has occurred with time as the sediment has built up. Near the surface, and in young sediments, the microflora is dominated by aerobic heterotrophs, but their activity rapidly depletes the available oxygen, and they are gradually replaced by anaerobic heterotrophs. These include bacteria which oxidize organic compounds by reducing SO_4^{--} ions to hydrogen sulphide, and this well-known indicator of anaerobic conditions accumulates in the interstitial water. In the narrow transition zone between the aerobic and anaerobic sediments, where both H_2S and O_2 are present, chemoautotrophic bacteria utilize the energy released by the oxidation of H_2S to elemental sulphur or SO_4^{--} in the synthesis of new organic compounds. Thus, they regenerate SO_4^{--} ions and add to the organic carbon available to the heterotrophs. Small pockets of anaerobic conditions may also develop within the aerobic zone (e.g. in the centre of faecal pellets), and the H_2S which is released may support a small population of chemoautotrophs throughout the upper part

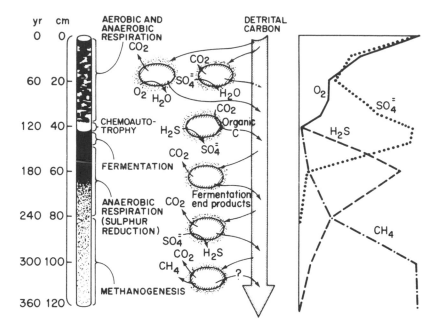

Fig. 9.2 Distribution of different types of microbial metabolism in a core of sediment from 24 m water depth in Halifax Harbour, Nova Scotia, and depth profiles of the main inorganic chemicals involved.[181]

of the sediments.[114] The sulphate concentration at each level is determined by the balance between chemosynthetic and sulphate-reducing bacteria, but the re-oxidation of H_2S is not possible deep in the anaerobic zone. Therefore, anaerobic methanogenic bacteria replace those dependent on sulphate, and methane becomes the dominant product of decomposition in old, deep sediments (Fig. 9.2). Green or purple sulphur bacteria, which utilize reduced sulphur compounds as electron donors in photosynthesis, may also occur if the anaerobic zone develops close to the surface of the sediments, or if the lower layers of a shallow water column become anaerobic. However, the appropriate combination of the presence of light and the absence of oxygen is rather rare in the sea, and marine photosynthetic bacteria are probably of little quantitative significance.

Ocean sediments that are in water over 1000 m deep have a low organic content and are subjected to high hydrostatic pressures, low temperatures and complete darkness (Table 9.1). Such environments are clearly rather difficult to investigate, and the first microbiological experiment to be carried out *in situ* was the result of an accident. In 1968, a deep-sea submersible, the *Alvin*, sank in 1540 m of water and, although the crew escaped, they left behind their lunch of sandwiches and apples in a plastic box. When the *Alvin* was recovered 11 months later, the lunch was still very well preserved, and this indicated that the physical conditions in deep water prevent rapid utilization of organic matter, even when there is plenty of it. Subsequent laboratory experiments with natural microbial populations in simulated deep-sea conditions have confirmed that bacteria grow very slowly at depth (Table 9.1), and have shown that the high pressures in deep water inhibit bacterial activity more strongly than the low temperatures. Even if elaborate sampling chambers are used, so that the natural populations are not decompressed as they are brought to the surface, their metabolic rates are only a fraction of those of similar populations incubated at the same deep-sea temperatures but under normal atmospheric pressure (Fig. 9.3). Thus, deep-water bacteria do not seem to have adapted to the high pressures of their natural environment, and energy flow in deep-sea ecosystems appears to be a very slow process.[104]

This conclusion may have to be modified in the light of the recent discovery of dense populations of large and unusual animals in the immediate vicinity of warm-water vents on the ocean floor.[43] Because of the limited amount of organic matter known to reach the deep ocean from the photic zone, it was at first suggested the these animals must depend on primary production by chemosynthetic sulphur bacteria within the vents. This would mean that these hydrothermal vent communities differ from all other known ecosystems in that they are not ultimately dependent on photosynthesis. Although dense bacterial populations have been measured in the water close to the vents, there is as yet little evidence that their activity under *in situ* pressures is any greater than that of deep-sea heterotrophic bacteria.[120] An alternative explanation of the apparently high productivity of these communities is that the discharge of warm water creates convection currents, which suck in water from a large area of the surrounding sea floor,

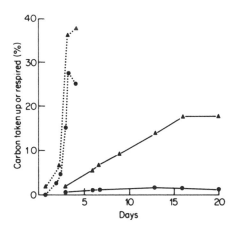

Fig. 9.3 Carbon uptake (●) and respiration (▲) of an undecompressed natural population of bacteria from a depth of 3000 m, measured at the *in situ* pressure (300 atm, continuous line) and at normal atmospheric pressure (1 atm, dotted line).[104] Both populations were incubated at 3.5°C. Vertical scale represents the percentage of labelled substrate supplied that was either incorporated into the bacterial cells, or released as $^{14}CO_2$.

and this concentrates organic detritus from the whole of this area into the region of the vent.[70] If, in addition, the metabolic rates of the animals are low because of the high pressures, it would still be possible for these communities to be largely supported by surface photosynthesis. Their discovery is so recent and so startling, however, that any new information is awaited with excitement.

References

1. ABBOTT, I.A. (1978). The uses of seaweed as food in Hawaii. *Economic Botany*, **32**, 409–12.
2. ALDERMAN, D.J. (1976). Fungal diseases of marine animals. In *Recent Advances in Aquatic Mycology*, JONES, E.B.G. (ed.), pp. 223–60. Elek Books, London.
3. ANDERSON, J.M. and BARRETT, J. (1979). Chlorophyll-protein complexes of brown algae: P700 reaction centre and light-harvesting complexes. *CIBA Foundation Symposium*, **61**, 81–96.
4. ANDREWS, J.H. (1979). Pathology of seaweeds: Current status and future prospects. *Experientia*, **35**, 429–50.
5. ANTIA, N.J., BERLAND, B.R. and BONIN, D.J. (1980). Proposal for an abridged nitrogen turnover cycle in certain marine planktonic systems involving hypoxanthine-guanine excretion by ciliates and their reutilization by phytoplankton. *Marine Ecology Progress Series*, **2**, 97–103.
6. ANTIA, N.J., BERLAND, B.R., BONIN, D.J. and MAESTRINI, S.Y. (1975). Comparative evaluation of certain organic and inorganic sources of nitrogen for phototrophic growth of marine microalgae. *Journal of the Marine Biological Association of the United Kingdom*, **55**, 519–39.
7. APPLEBY, G., COLBECK, J., HOLDSWORTH, E.S. and WADMAN, H. (1980). β-carboxylation enzymes in marine phytoplankton and isolation and purification of pyruvate carboxylase from *Amphidinium carterae* (Dinophyceae). *Journal of Phycology*, **16**, 290–95.
8. BAARDSETH, E. (1970). A square-scanning, two-stage sampling method of estimating seaweed quantities. *Norwegian Institute for Seaweed Research Report*, **33**, 1–41.
9. BAARDSETH, E. (1970). Synopsis of biological data on knobbed wrack *Ascophyllum nodosum* (Linnaeus) Le Jolis. *F.A.O. Fisheries Synopsis*, **38**, 40 pp.
10. BARDACH, J.E., RYTHER, J.H. and McLARNEY, W.O. (1972). *Aquaculture: the farming and husbandry of freshwater and marine organisms*. Wiley, New York.
11. BEARDALL, J. and MORRIS, I. (1976). The concept of light intensity adaptation in marine phytoplankton; some experiments with *Phaeodactylum tricornutum*. *Marine Biology*, **37**, 377–87.
12. BEARDALL, J., MUKERJI, D., GLOVER, H. and MORRIS, I. (1976). The path of carbon in photosynthesis by marine phytoplankton. *Journal of Phycology*, **12**, 409–17.
13. BENEDICT, C.R., WONG, W.W.L. and WONG, J.H.H. (1980). Fractionation of the stable isotopes of inorganic carbon by seagrasses. *Plant Physiology, Lancaster*, **65**, 512–17.
14. BENTRUP, F.W. (1963). Vergleichende Untersuchungen zur Polaritätsinduktion

durch das Licht an der *Equisetum*-spore und der *Fucus*-zygote. *Planta*, **59**, 472 – 91.

15. BERMAN, T. and HOLM-HANSEN, O. (1974). Release of photoassimilated carbon as dissolved organic matter by marine phytoplankton. *Marine Biology*, **28**, 305 – 10.

16. BIRMINGHAM, B.C. and COLMAN, B. (1979). Measurement of carbon dioxide compensation points of freshwater algae. *Plant Physiology, Lancaster*, **64**, 892 – 95.

17. BOADEN, P.J.S. and DRING, M.J. (1980). A quantitative evaluation of the effects of *Ascophyllum* harvesting on the littoral ecosystem. *Helgoländer Meeresuntersuchungen*, **33**, 700 – 10.

18. BOARDMAN, N.K. (1977). Comparative photosynthesis of sun and shade plants. *Annual Review of Plant Physiology*, **28**, 355 – 77.

19. BOLD, H.C. and WYNNE, M.J. (1978). *Introduction to the Algae*. Prentice-Hall, New Jersey.

20. BOLWELL, G.P., CALLOW, J.A., CALLOW, M.E. and EVANS, L.V. (1977). Cross-fertilization in fucoid seaweeds. *Nature, London*, **268**, 626 – 7.

21. BOOTH, E. (1975). Seaweeds in industry. In *Chemical Oceanography*, RILEY, J.P. and SKIRROW, G. (eds), Vol. 4, 2nd edn, pp. 219 – 68. Academic Press, London.

22. BRIGGS, J.C. (1974). *Marine Zoogeography*. McGraw-Hill, New York.

23. BRINKHUIS, B.H. (1977). Comparisons of salt-marsh fucoid production estimated from three different indices. *Journal of Phycology*, **13**, 328 – 35.

24. BROWN, T.E. and RICHARDSON, F.L. (1968). The effect of growth environment on the physiology of algae: light intensity. *Journal of Phycology*, **4**, 38 – 54.

25. BUGGELN, R.G. (1974). Negative phototropism of the haptera of *Alaria esculenta* (Laminariales). *Journal of Phycology*, **10**, 80 – 2.

26. BUTLER, E.I., KNOX, S. and LIDDICOAT, M.I. (1979). The relationship between inorganic and organic nutrients in sea water. *Journal of the Marine Biological Association of the United Kingdom*, **59**, 239 – 50.

27. CAREFOOT, T. (1977). *Pacific Seashores*. University of Washington Press, Seattle.

28. CARLUCCI, A.F. and WILLIAMS, P.M. (1978). Simulated *in situ* growth rates of pelagic marine bacteria. *Naturwissenschaften*, **65**, 541 – 42.

29. CHAPMAN, A.R.O. (1974). The ecology of macroscopic marine algae. *Annual Review of Ecology and Systematics*, **5**, 65 – 80.

30. CHAPMAN, A.R.O. and CRAIGIE, J.S. (1977). Seasonal growth in *Laminaria longicruris*: relations with dissolved inorganic nutrients and internal reserves of nitrogen. *Marine Biology*, **40**, 197 – 205.

31. CHAPMAN, A.R.O. and CRAIGIE, J.S. (1978). Seasonal growth in *Laminaria longicruris*: relations with reserve carbohydrate storage and production. *Marine Biology*, **46**, 209 – 13.

32. CHAPMAN, V.J. (1970). *Seaweeds and their Uses*. Methuen, London.

33. CHAPMAN, V.J. and CHAPMAN, D.J. (1973). *The Algae*, 2nd edn. MacMillan, London.

34. CHAPMAN, V.J. and CHAPMAN, D.J. (1976). Life forms in the algae. *Botanica Marina*, **19**, 65 – 74.

35. CHARTERS, A.C., NEUSHUL, M. and COON, D. (1973). The effect of water motion on algal spore adhesion. *Limnology and Oceanography*, **18**, 884 – 96.

36. CHENG, T.-H. (1969). Production of kelp — a major aspect of China's exploitation of the sea. *Economic Botany*, **23**, 215 – 36.

37. CHRISTIE, A.O., EVANS, L.V. and SHAW, M. (1970). Studies on the ship-fouling alga *Enteromorpha*. II. The effect of certain enzymes on the adhesion of zoospores. *Annals of Botany*, **34**, 467 – 82.

38. CLAYTON, R.K. (1980). *Photosynthesis: physical mechanisms and chemical patterns*. University Press, Cambridge.

39. CLENDENNING, K.A. (1971). Organic productivity in kelp areas. *Beihefte zur Nova Hedwigia*, **32**, 259 – 63.

40. CODD, G.A. and STEWART, R. (1980). The photoinactivation of micro-algal ribulose

bisphosphate carboxylase: its physiological and ecological significance. In *The Blue Light Syndrome*, SENGER, H. (ed.), pp. 392–400. Springer-Verlag, Berlin.

41. CONOVER, J.T. (1968). The importance of natural diffusion gradients and transport of substances related to benthic marine plant metabolism. *Botanica Marina*, **11**, 1–9.

42. CONWAY, H.L., HARRISON, P.J. and DAVIS, C.O. (1976). Marine diatoms grown in chemostats under silicate or ammonium limitation. II. Transient response of *Skeletonema costatum* to a single addition of the limiting nutrient. *Marine Biology*, **35**, 187–99.

43. CORLISS, J.B. *et al.* (1979). Submarine thermal springs on the Galápagos Rift. *Science, N.Y.*, **203**, 1073–83.

44. COUGHLAN, S. (1977). Sulphate uptake in *Fucus serratus*. *Journal of Experimental Botany*, **28**, 1207–15.

45. COUGHLAN, S. and TATTERSFIELD, D. (1977). Photorespiration in larger littoral algae. *Botanica Marina*, **20**, 265–6.

46. CRAIGIE, J.S. (1974). Storage products. In *Algal Physiology and Biochemistry*, STEWART, W.D.P., pp. 206–35. Blackwell Scientific Publications, Oxford.

47. CRAYTON, M.A., WILSON, E. and QUATRANO, R.S. (1974). Sulphation of fucoidan in *Fucus* embryos. II. Separation from initiation of polar growth. *Developmental Biology*, **39**, 164–7.

48. CUNDELL, A.M., SLEETER, T.D. and MITCHELL, R. (1977). Microbial populations associated with the surface of the brown alga *Ascophyllum nodosum*. *Microbial Ecology*, **4**, 81–91.

49. CUSHING, D.H. (1975). *Marine Ecology and Fisheries*. University Press, Cambridge.

50. DALBY, D.H., COWELL, E.B. SYRATT, W.J. and CROTHERS, J.H. (1978). An exposure scale for marine shores in western Norway. *Journal of the Marine Biological Association of the United Kingdom*, **58**, 975–96.

51. DAVID, H.M. (1943). Studies in the autecology of *Ascophyllum nodosum* Le Jol. *Journal of Ecology*, **31**, 178–98.

52. DAYTON, P.K. (1973). Dispersion, dispersal and persistence of the annual intertidal alga, *Postelsia palmaeformis* Ruprecht. *Ecology*, **54**, 433–8.

53. DAYTON, P.K. (1975). Experimental evaluation of ecological dominance in a rocky intertidal algal community. *Ecological Monographs*, **45**, 137–59.

54. DIXON, P.S. (1971). Cell enlargement in relation to the development of thallus form in Florideophyceae. *British Phycological Journal*, **6**, 195–205.

55. DOTY, M.S. (1971). Measurement of water movement in reference to benthic algal growth. *Botanica Marina*, **14**, 32–5.

56. DOTY, M.S. (1979). Status of marine agronomy, with special reference to the tropics. *Proceedings of the International Seaweed Symposia*, **9**, 35–58.

57. DREW, E.A. (1978). Factors affecting photosynthesis and its seasonal variation in the seagrasses *Cymodocea nodosa* (Ucria) Aschers, and *Posidonia oceanica* (L.) Delile in the Mediterranean. *Journal of Experimental Marine Biology and Ecology*, **31**, 173–94.

58. DREW, K.M. (1949). *Conchocelis*-phase in the life-history of *Porphyra umbilicalis* (L.) Kütz. *Nature, London*, **164**, 748–9.

59. DRING, M.J. (1970). Photoperiodic effects in microorganisms. In *Photobiology of Microorganisms*, HALLDAL, P. (ed.), pp. 345–68. Wiley, London.

60. DRING, M.J. (1981). Chromatic adaptation of photosynthesis in benthic marine algae: An examination of its ecological significance using a theoretical model. *Limnology and Oceanography*, **26**, 271–84.

61. DRING, M.J. (1981). Photosynthesis and development of marine macrophytes in natural light spectra. In *Plants and the Daylight Spectrum*, SMITH, H. (ed.), pp. 297–314. Academic Press, London.

62. DRING, M.J. and LÜNING, K. (1975). Induction of two-dimensional growth and hair formation by blue light in the brown alga *Scytosiphon lomentaria*. *Zeitschrift für Pflanzenphysiologie*, **75**, 107–17.

63. DRING, M.J. and LÜNING, K. (1975). A photoperiodic response mediated by blue light in the brown alga *Scytosiphon lomentaria*. *Planta*, **125**, 25 – 32.
64. DROMGOOLE, F.I. (1980). Desiccation resistance of intertidal and subtidal algae. *Botanica Marina*, **23**, 149 – 59.
65. DRUEHL, L.D. (1970). The pattern of Laminariales distribution in the northeast Pacific. *Phycologia*, **9**, 237 – 47.
66. DUFFIELD, E.C.S., WAALAND, S.D. and CLELAND, R. (1972). Morphogenesis in the red alga, *Griffithsia pacifica*: regeneration from single cells. *Planta*, **105**, 185 – 95.
67. DURBIN, E.G. (1978). Aspects of the biology of resting spores of *Thalassiosira nordenskioeldii* and *Detonula confervacea*. *Marine Biology*, **45**, 31 – 7.
68. EKMAN, S. (1953). *Zoogeography of the Sea*. Sidgwick & Jackson, London.
69. ENGELMANN, T.W. (1883). Farbe und Assimilation. *Botanische Zeitung*, **41**, 1 – 29.
70. ENRIGHT, J.T., NEWMAN, W.A., HESSLER, R.R. and McGOWAN, J.A. (1981). Deep-ocean hydrothermal vent communities. *Nature, London*, **289**, 219 – 21.
71. EPPLEY, R.W. (1980) Primary productivity in the sea. *Nature, London*, **286**, 109 – 10.
72. EPPLEY, R.W., ROGERS, J.N. and McCARTHY, J.J. (1969). Half-saturation constants for uptake of nitrate and ammonium by marine phytoplankton. *Limnology and Oceanography*, **14**, 912 – 20.
73. EPPLEY, R.W. and THOMAS, W.H. (1969). Comparison of half-saturation constants for growth and nitrate uptake of marine phytoplankton. *Journal of Phycology*, **5**, 375 – 9.
74. EVANS, L.V. (1965). Cytological studies in the Laminariales. *Annals of Botany*, **29**, 541 – 62.
75. FAUST, M.A. and GANTT, E. (1973). Effect of light intensity and glycerol on the growth, pigment composition and ultrastructure of *Chroomonas* sp. *Journal of Phycology*, **9**, 489 – 95.
76. FINDENEGG, G.R. (1980). Inorganic carbon transport in microalgae. II. Uptake of HCO_3^- ions during photosynthesis of five microalgal species. *Plant Science Letters*, **18**, 289 – 97.
77. FJELD, A. and LØVLIE, A. (1976). Genetics of multicellular marine algae. In *The Genetics of Algae*, LEWIN, R.A. (ed.), pp. 219 – 35. Blackwell Scientific Publications, Oxford.
78. FORTES, M.D. and LÜNING, K. (1980). Growth rates of North Sea macroalgae in relation to temperature, irradiance and photoperiod. *Helgoländer Meeresuntersuchungen*, **34**, 15 – 29.
79. FRIES, L. (1966). Influence of iodine and bromine on growth of some red algae in axenic culture. *Physiologia Plantarum*, **19**, 800 – 8
80. FRIES, L. (1975). Requirement of bromine in a red alga. *Zeitschrift für Pflanzenphysiologie*, **76**, 366 – 8.
81. GANTT, E. (1981). Phycobilisomes. *Annual Review of Plant Physiology*, **32**, 327 – 47.
82. GARBARY, D.J., GRUND, D. and McLACHLAN, J. (1978). The taxonomic status of *Ceramium rubrum* (Huds.) C. Ag. (Ceramiales, Rhodophyceae) based on culture experiments. *Phycologia*, **17**, 85 – 94.
83. GLOVER, H.E. and MORRIS, I. (1979). Photosynthetic carboxylating enzymes in marine phytoplankton. *Limnology and Oceanography*, **24**, 510 – 9.
84. GOLDMAN, J.C., McCARTHY, J.J. and PEAVEY, D.G. (1979). Growth rate influence on the chemical composition of phytoplankton in oceanic waters. *Nature, London*, **279**, 210 – 5.
85. GRIME, J.P. (1979). *Plant Strategies and Vegetation Processes*. Wiley, Chichester.
86. HALL, C.A.S. and MOLL, R. (1975). Methods of assessing aquatic primary productivity. In *Primary Productivity of the Biosphere*, LIETH, H. and WHITTAKER, R.H. (eds), pp. 19 – 53. Springer-Verlag, Berlin.
87. HARRIS, G.P. (1980). The measurement of photosynthesis in natural populations of phytoplankton. In *The Physiological Ecology of Phytoplankton*, MORRIS, I. (ed.), pp. 129 – 87. Blackwell Scientific Publications, Oxford.

88. HARRISON, P.G. and MANN, K.H. (1975). Detritus formation from eelgrass (*Zostera marina* L.). The relative effects of fragmentation, leaching and decay. *Limnology and Oceanography*, **20**, 924 – 34.
89. HARTOG, C. den (1970). *The Sea-grasses of the World*. North-Holland Publishing Company, Amsterdam.
90. HASEGAWA, Y. (1976). Progress of *Laminaria* cultivation in Japan. *Journal of the Fisheries Research Board of Canada*, **33**, 1002 – 6.
91. HASLE, G.R. and HEIMDAL, B.R. (1968). Morphology and distribution of the marine centric diatom *Thalassiosira antarctica* Comber. *Journal of the Royal Microscopical Society*, **88**, 357 – 69.
92. HATCHER, B.G., CHAPMAN, A.R.O. and MANN, K.H. (1977). An annual carbon budget for the kelp *Laminaria longicruris*. *Marine Biology*, **44**, 85 – 96.
93. HAWKINS, S.J. (1981). The influence of season and barnacles on the algal colonization of *Patella vulgata* exclusion areas. *Journal of the Marine Biological Association of the United Kingdom*, **61**, 1 – 15.
94. HELLEBUST, J.A., TERBORGH, J. and McLEOD, G.C. (1967). The photosynthetic rhythm of *Acetabularia crenulata*. II. Measurements of photoassimilation of carbon dioxide and the activities of enzymes of the reductive pentose cycle. *Biological Bulletin*, **133**, 670 – 78.
95. HOEK, C. van den (1975). Phytogeographic provinces along the coasts of the northern Atlantic Ocean. *Phycologia*, **14**, 317 – 30.
96. HOEK, C. van den, and DONZE, M. (1967). Algal phytogeography of the European Atlantic coasts. *Blumea*, **15**, 63 – 89.
97. HOLDSWORTH, E.S. and ARSHAD, J.H. (1977). A manganese-copper-pigment-protein complex isolated from the photosystem II of *Phaeodactylum tricornutum*. *Archives of Biochemistry and Biophysics*, **183**, 361 – 73.
98. HOPPE, H.-G. (1978). Relations between active bacteria and heterotrophic potential in the sea. *Netherlands Journal of Sea Research*, **12**, 78 – 98.
99. HUGHES, G.C. (1974). Geographical distribution of the higher marine fungi. *Veroffentlichungen des Instituts für Meeresforschung, Bremerhaven, Supplement*, **5**, 419 – 41.
100. IGNATIADES, L. and FOGG, G.E. (1973). Studies on the factors affecting the release of organic matter by *Skeletonema costatum* (Greville) Cleve in culture. *Journal of the Marine Biological Association of the United Kingdom*, **53**, 937 – 56.
101. JACKSON, P. (1957). Harvesting machinery for brown sublittoral seaweeds. *Engineer*, **203**, 400 – 2 and 439 – 41.
102. JAENICKE, L. (1977). Sex hormones of brown algae. *Naturwissenschaften*, **64**, 69 – 75.
103. JAFFE, L.F. (1968). Localization in the developing *Fucus* egg and the general role of localizing currents. *Advances in Morphogenesis*, **7**, 295 – 328.
104. JANNASCH, H.W. and WIRSEN, C.O. (1977). Microbial life in the deep sea. *Scientific American*, **236 (6)**, 42 – 52.
105. JEFFREY, S.W. (1976). The occurrence of chlorophyll c_1 and c_2 in algae. *Journal of Phycology*, **12**, 349 – 54.
106. JENSEN, A. (1979). Industrial utilization of seaweeds in the past, present and future. *Proceedings of the International Seaweed Symposia*, **9**, 17 – 34.
107. JERLOV, N.G. (1976). *Marine Optics*. Elsevier Scientific Publishing Co., Amsterdam.
108. JITTS, H.R., McALLISTER, C.D., STEPHENS, K. and STRICKLAND, J.D.H. (1964). The cell division rates of some marine phytoplankters as a function of light and temperature. *Journal of the Fisheries Research Board of Canada*, **21**, 139 – 57.
109. JOHNSON, W.S., GIGON, A., GULMON, S.L. and MOONEY, H.A. (1974). Comparative photosynthetic capacities of intertidal algae under exposed and submerged conditions. *Ecology*, **55**, 450 – 3.
110. JOLIFFE, E.A. and TREGUNNA, E.B. (1970). Studies on HCO_3^- ion uptake during photosynthesis in benthic marine algae. *Phycologia*, **9**, 293 – 303.
111. JONES, N.S. and KAIN, J.M. (1967). Subtidal algal colonization following the removal of *Echinus*. *Helgoländer Meeresuntersuchungen*, **15**, 460 – 6.

112. JONES, W.E. and DEMETROPOULOS, A. (1968). Exposure to wave action: measurements of an important ecological parameter on rocky shores in Anglesey. *Journal of Experimental Marine Biology and Ecology*, **2**, 46–63.
113. JORDAN, A.J. and VADAS, R.L. (1972). Influence of environmental parameters on intraspecific variation in *Fucus vesiculosus*. *Marine Biology*, **14**, 248–52.
114. JØRGENSEN, B.B. (1977). Bacterial sulphate reduction within reduced microniches of oxidized marine sediments. *Marine Biology*, **41**, 7–17.
115. JØRGENSEN, E.G. (1968). The adaptation of plankton algae. II. Aspects of the temperature adaptation of *Skeletonema costatum*. *Physiologia Plantarum*, **21**, 423–7.
116. JØRGENSEN, E.G. (1969). The adaptation of plankton algae. IV. Light adaptation in different algal species. *Physiologia Plantarum*, **22**, 1307–15.
117. KAIN, J.M. (1966). The role of light in the ecology of *Laminaria hyperborea*. In *Light as an Ecological Factor*, BAINBRIDGE, R., EVANS, G.C. and RACKHAM, O. (eds), pp. 319–34. Blackwell Scientific Publications, Oxford.
118. KAIN, J.M. (1977). The biology of *Laminaria hyperborea*. X. The effect of depth on some populations. *Journal of the Marine Biological Association of the United Kingdom*, **57**, 587–607.
119. KAIN, J.M. (1979). A view of the genus *Laminaria*. *Oceanography and Marine Biology, Annual Review*, **17**, 101–61.
120. KARL, D.M., WIRSEN, C.O. and JANNASCH, H.W. (1980). Deep-sea primary production at the Galápagos hydrothermal vents. *Science, N.Y.*, **207**, 1345–7.
121. KERSHAW, K.A. (1973). *Quantitative and Dynamic Plant Ecology*, 2nd edn. Edward Arnold, London.
122. KILHAM, P. and KILHAM, S.S. (1980). The evolutionary ecology of phytoplankton. In *The Physiological Ecology of Phytoplankton*, MORRIS, I. (ed.), pp. 571–97. Blackwell Scientific Publications, Oxford.
123. KLIKOFF, L.G. (1966). Temperature dependence of the oxidative rates of mitochondria in *Danthonia intermedia*, *Penstemon davidsonii* and *Sitanion hystrix*. *Nature, London*, **212**, 529–30.
124. KOCHERT, G. (1978). Sexual pheromones in algae and fungi. *Annual Review of Plant Physiology*, **29**, 461–86.
125. KOHLMEYER, J. and KOHLMEYER, E. (1979). *Marine Mycology: The Higher Fungi*. Academic Press, New York.
126. KORNMANN, P. and SAHLING, P.-H. (1974). Prasiolales (Chlorophyta) von Helgoland. *Helgoländer Meeresuntersuchungen*, **26**, 99–133.
127. KREMER, B.P. and BERKS, R. (1978). Photosynthesis and carbon metabolism in marine and freshwater diatoms. *Zeitschrift für Pflanzenphysiologie*, **87**, 149–65.
128. KREMER, B.P. and KÜPPERS, U. (1977). Carboxylating enzymes and pathway of photosynthetic carbon assimilation in different marine algae — evidence for the C_4-pathway? *Planta*, **133**, 191–6.
129. KRISTENSEN, I. (1968). Surf influence on the thallus of fucoids and the rate of desiccation. *Sarsia*, **34**, 69–82.
130. KÜPPERS, U. and KREMER, B.P. (1978). Longitudinal profiles of carbon dioxide fixation capacities in marine macroalgae. *Plant Physiology, Lancaster*, **62**, 49–53.
131. LAYCOCK, R.A. (1974). The detrital food chain based on seaweeds. I. Bacteria associated with the surface of *Laminaria* fronds. *Marine Biology*, **25**, 223–31.
132. LEVITT, J. (1975). *Responses of Plants to Environmental Stresses*. Academic Press, New York.
133. LEWIN, J. (1966). Boron as a growth requirement for diatoms. *Journal of Phycology*, **2**, 160–3.
134. LEWIN, R.A. (1976). *The Genetics of Algae*. Blackwell Scientific Publications, Oxford.
135. LEWIS, J.R. (1964). *The Ecology of Rocky Shores*. English Universities Press, London.

136. LITTLER, M.M. and LITTLER, D.S. (1980). The evolution of thallus form and survival strategies in benthic marine macroalgae: field and laboratory tests of a functional form model. *American Naturalist*, **116**, 25 – 44.

137. LLOYD, N.D.H., CANVIN, D.T. and CULVER, D.A. (1977). Photosynthesis and photorespiration in algae. *Plant Physiology, Lancaster*, **59**, 936 – 40.

138. LUBCHENCO, J. and MENGE, B.A. (1978). Community development and persistence in a low rocky intertidal zone. *Ecological Monographs*, **48**, 67 – 94.

139. LUND, J.W.G., MACKERETH, F.J.H. and MORTIMER, C.H. (1963). Changes in depth and time of certain chemical and physical conditions and of the standing crop of *Asterionella formosa* Hass. in the North Basin of Windemere in 1947. *Philosophical Transactions of the Royal Society*, B, **246**, 255 – 90.

140. LÜNING, K. (1971). Seasonal growth of *Laminaria hyperborea* under recorded underwater light conditions near Helgoland. In *Fourth European Marine Biology Symposium*, CRISP, D.J. (ed.), pp. 347 – 61. University Press, Cambridge.

141. LÜNING, K. (1975). Kreuzungsexperimente an *Laminaria saccharina* von Helgoland und von der Isle of Man. *Helgoländer Meeresuntersuchungen*, **27**, 108 – 14.

142. LÜNING, K. (1980). Critical levels of light and temperature regulating the gametogenesis of three *Laminaria* species (Phaeophyceae). *Journal of Phycology*, **16**, 1 – 15.

143. LÜNING, K. (1980). Control of algal life history by daylength and temperature. In *The Shore Environment*, Vol. 2: *Ecosystems*, PRICE, J.H., IRVINE, D.E.G. and FARNHAM, W.F. (eds), pp. 915 – 45. Academic Press, London.

144. LÜNING, K. (1981). Light. In *The Biology of Seaweeds*, LOBBAN, C.S. and WYNNE, M.J. (eds), pp. 326 – 55. Blackwell Scientific Publications, Oxford.

145. LÜNING, K. and DRING, M.J. (1975). Reproduction, growth and photosynthesis of gametophytes of *Laminaria saccharina* grown in blue and red light. *Marine Biology*, **29**, 195 – 200.

146. LÜNING, K. and DRING, M.J. (1979). Continuous underwater light measurement near Helgoland (North Sea) and its significance for characteristic light limits in the sublittoral region. *Helgoländer Meeresuntersuchungen*, **32**, 403 – 24.

147. LÜNING, K. and MÜLLER, D.G. (1978). Chemical interaction in sexual reproduction of several Laminariales (Phaeophyceae): Release and attraction of spermatozoids. *Zeitschrift für Pflanzenphysiologie*, **89**, 333 – 41.

148. MACKIE, W. and PRESTON, R.D. (1974). Cell wall and intercellular region polysaccharides. In *Algal Physiology and Biochemistry*, STEWART, W.D.P., pp. 40 – 85. Blackwell Scientific Publications, Oxford.

149. MacISAAC, J.J. and DUGDALE, R.C. (1969). The kinetics of nitrate and ammonia uptake by natural populations of marine phytoplankton. *Deep-Sea Research*, **16**, 45 – 57.

150. MacROBBIE, E.A.C. (1974). Ion Uptake. In *Algal Physiology and Biochemistry*, STEWART, W.D.P., pp. 676 – 713. Blackwell Scientific Publications, Oxford.

151. MADDUX, W.S. and JONES, R.F. (1964). Some interactions of temperature, light intensity, and nutrient concentration during the continuous culture of *Nitzschia closterium* and *Tetraselmis* sp. *Limnology and Oceanography*, **9**, 79 – 86.

152. MANDELLI, E.F. (1972). The effect of growth illumination on the pigmentation of a marine dinoflagellate. *Journal of Phycology*, **8**, 367 – 9.

153. MANN, K.H. (1973). Seaweeds: their productivity and strategy for growth. *Science, N.Y.* **182**, 975 – 81.

154. McCARTHY, J.J. and GOLDMAN, J.C. (1979). Nitrogenous nutrition of marine phytoplankton in nutrient-depleted waters. *Science, N.Y.* **203**, 670 – 2.

155. McINTYRE, A. (1967). Coccoliths as palaeoclimatic indicators of Pleistocene glaciation. *Science, N.Y.*, **158**, 1314 – 7.

156. McINTYRE, A. and BÉ, A.W.H. (1967). Modern Coccolithophoridae of the Atlantic Ocean — I. Placoliths and cyrtoliths. *Deep-Sea Research*, **14**, 561 – 97.

157. McLACHLAN, J. (1977). Effects of nutrients on growth and development of

embryos of *Fucus edentatus* Pyl. (Phaeophyceae, Fucales). *Phycologia*, **16**, 329–38.

158. McROY, C.P. and McMILLAN, C. (1977). Production ecology and physiology of seagrasses. In *Seagrass Ecosystems*, McROY, C.P. and HELFFERICH, C. (eds), pp. 53–87. Dekker, New York.

159. MEER, J.P. van der (1978). Genetics of *Gracilaria* sp. (Rhodophyceae, Gigartinales). III. Non-Mendelian gene transmission. *Phycologia*, **17**, 314–18.

160. MEER, J.P. van der, and TODD, E.R. (1977). Genetics of *Gracilaria* sp. (Rhodophyceae, Gigartinales). IV. Mitotic recombination and its relationship to mixed phases in the life history. *Canadian Journal of Botany*, **55**, 2810–7.

161. MOHR, H. (1972). *Lectures on Photomorphogenesis*. Springer-Verlag, Berlin.

162. MOORE, H.B. (1935). The biology of *Balanus balanoides*. IV. Relation to environmental factors. *Journal of the Marine Biological Association of the United Kingdom*, **20**, 279–307.

163. MORRIS, I. and FARRELL, K. (1971). Photosynthetic rates, gross patterns of carbon dioxide assimilation and activities of ribulose diphosphate carboxylase in marine algae grown at different temperatures. *Physiologia Plantarum*, **25**, 372–7.

164. MORRIS, I. and GLOVER, H.E. (1974). Questions on the mechanism of temperature adaptation in marine phytoplankton. *Marine Biology*, **24**, 147–54.

165. MOSS, B. and SHEADER, A. (1973). The effect of light and temperature upon the germination and growth of *Halidrys siliquosa* (L.) Lyngb. (Phaeophyceae, Fucales). *Phycologia*, **12**, 63–8.

166. MUKERJI, D., GLOVER, H.E. and MORRIS, I. (1978). Diversity in the mechanism of carbon dioxide fixation in *Dunaliella tertiolecta* (Chlorophyceae). *Journal of Phycology*, **14**, 137–42.

167. MÜLLER, D. (1962). Uber jahres- und lunarperiodische Erscheinungen bei einigen Braunalgen. *Botanica Marina*, **4**, 140–55.

168. MÜLLER, D.G. (1972). Studies on reproduction in *Ectocarpus siliculosus*. *Société Botanique de France, Mémoires*, 1972, 87–98.

169. MÜLLER, D.G. (1976). Quantitative evaluation of sexual chemotaxis in two marine brown algae. *Zeitschrift für Pflanzenphysiologie*, **80**, 120–30.

170. MÜLLER, D.G. and GASSMANN, G. (1978). Identification of the sex attractant in the marine brown alga *Fucus vesiculosus*. *Naturwissenschaften*, **65**, 389–90.

171. MÜLLER, D.G. and SEFERIADIS, K. (1977). Specificity of sexual chemotaxis in *Fucus serratus* and *Fucus vesiculosus* (Phaeophyceae). *Zeitschrift für Pflanzenphysiologie*, **84**, 85–94.

172. MURRAY, S.N. and DIXON, P.S. (1973). The effect of light intensity and light period on the development of thallus form in the marine red alga *Pleonosporium squarrulosum* (Harvey) Abbott (Rhodophyta: Ceramiales). I. Apical cell division — main axes. *Journal of Experimental Marine Biology and Ecology*, **13**, 15–27.

173. MUUS, B.J. (1968). A field method for measuring 'exposure' by means of plaster balls. *Sarsia*, **34**, 61–8.

174. NAKAMURA, K., OGAWA, T. and SHIBATA, K. (1976). Chlorophyll and peptide compositions in the two photosystems of marine green algae. *Biochimica et Biophysica Acta*, **423**, 227–36.

175. NEAME, K.D. and RICHARDS, T.G. (1972). *Elementary Kinetics of Membrane Carrier Transport*. Blackwell Scientific Publications, Oxford.

176. NEILSON, A.H. and LARSSON, T. (1980). The utilization of organic nitrogen for growth of algae: physiological aspects. *Physiologia Plantarum*, **48**, 542–53.

177. NEISH, A.C., SHACKLOCK, P.F., FOX, C.H. and SIMPSON, F.J. (1977). The cultivation of *Chondrus crispus*. Factors affecting growth under greenhouse conditions. *Canadian Journal of Botany*, **55**, 2263–71.

178. NEUMANN, R. (1979). Bacterial induction of settlement and metamorphosis in the planula larvae of *Cassiopeia andromeda* (Cnidaria: Scyphozoa, Rhizostomeae). *Marine Ecology Progress Series*, **1**, 21–8.

179. NORTH, W.J. (1977). Possibilities of biomass from the ocean. The Marine Farm

Project. In *Biological Solar Energy Conversion*, MITSUI, A., MIYACHI, S., SAN PIETRO, A. and TAMURA, S. (eds), pp. 347 – 61. Academic Press, New York.

180. NORTON, T.A. (1977). Experiments on the factors influencing the geographical distributions of *Saccorhiza polyschides* and *Saccorhiza dermatodea*. *New Phytologist*, **78**, 625 – 35.

181. NOVITSKY, J.A. and KEPKAY, P.E. (1981). Patterns of microbial heterotrophy through changing environments in a marine sediment. *Marine Ecology Progress Series*, **4**, 1 – 7.

182. OGATA, E. and MATSUI, T. (1965). Photosynthesis in several marine plants of Japan as affected by salinity, drying and pH, with attention to their growth habitats. *Botanica Marina*, **7/8**, 199 – 217.

183. O'KELLEY, J.C. (1974). Inorganic nutrients. In *Algal Physiology and Biochemistry*, STEWART, W.D.P., pp. 610 – 35, Blackwell Scientific Publications, Oxford.

184. OLTMANNS, F. (1892). Über die Kultur and Lebensbedingungen der Meeresalgen. *Jahresbericht der Wissenschaftlichen Botanik*, **23**, 349 – 440.

185. PAINE, R.T. (1979). Disaster, catastrophe, and local persistence of the sea palm *Postelsia palmaeformis*. *Science, N. Y.*, **205**, 685 – 7.

186. PAPENFUSS, G.F. (1972). On the geographical distribution of some tropical marine algae. *Proceedings of the International Seaweed Symposia*, **7**, 45 – 51.

187. PEARSE, J.S. and HINES, A.H. (1979). Expansion of a central Californian kelp forest following the mass mortality of sea urchins. *Marine Biology*, **51**, 83 – 91.

188. PEDERSEN, M. and FRIDBORG, G. (1972). Cytokinin-like activity in sea water from the *Fucus-Ascophyllum* zone. *Experientia*, **28**, 111.

189. PERCIVAL, E. (1979). The polysaccharides of green, red and brown seaweeds: their basic structure, biosynthesis and function. *British Phycological Journal*, **14**, 103 – 17.

190. PIELOU, E.C. (1978). Latitudinal overlap of seaweed species: evidence for quasi-sympatric speciation. *Journal of Biogeography*, **5**, 227 – 38.

191. PLATT, T. and JASSBY, A.D. (1976). The relationship between photosynthesis and light for natural assemblages of coastal marine phytoplankton. *Journal of Phycology*, **12**, 424 – 30.

192. POMEROY, L.R. (1980). Detritus and its role as a food source. In *Fundamentals of Aquatic Ecosystems*, BARNES, R.K. and MANN, K.H. (eds), pp. 84 – 102. Blackwell Scientific Publications, Oxford.

193. PRÉZELIN, B.B. (1976), The role of peridinin-chlorophyll *a*-proteins in the photosynthetic light adaptation of the marine dinoflagellate, *Glenodinium* sp. *Planta*, **130**, 225 – 33.

194. PRÉZELIN, B.B. and ALBERTE, R.S. (1978). Photosynthetic characteristics and organization of chlorophyll in marine dinoflagellates. *Proceedings of the National Academy of Science*, **75**, 1801 – 4.

195. PRÉZELIN, B.B., LEY, A.C. and HAXO, F.T. (1976). Effects of growth irradiance on the photosynthetic action spectra of the marine dinoflagellate, *Glenodinium* sp. Planta, **130**, 251 – 6.

196. PRÉZELIN, B.B. and SWEENEY, B.M. (1978). Photoadaptation of photosynthesis in *Gonyaulax polyedra*. *Marine Biology*, **48**, 27 – 35.

197. PRINGLE, J.D. (1979). Aspects of the ecological impact of *Chondrus crispus* (Florideophyceae) harvesting in eastern Canada. *Proceedings of the International Seaweed Symposia*, **9**, 225 – 32.

198. PROVASOLI, L. and CARLUCCI, A.F. (1974). Vitamins and growth regulators. In *Algal Physiology and Biochemistry*, STEWART, W.D.P., pp. 741 – 87. Blackwell Scientific Publications, Oxford.

199. PROVASOLI, L. and PINTNER, I.J. (1980). Bacteria induced polymorphism in an axenic laboratory strain of *Ulva lactuca* (Chlorophyceae). *Journal of Phycology*, **16**, 196 – 201.

200. PUISEUX-DAO, S. (1970). *Acetabularia and Cell Biology*. Logos Press, London.

201. QUADIR, A., HARRISON, P.J. and DeWREEDE, R.E. (1979). The effects of emergence and submergence on the photosynthesis and respiration of marine macrophytes. *Phycologia*, **18**, 83 – 8.

202. QUATRANO, R.S. (1978). Development of cell polarity. *Annual Review of Plant Physiology*, **29**, 487–510.
203. RAMUS, J. (1981). The capture and transduction of light energy. In *The Biology of Seaweeds*, LOBBAN, C.S. and WYNNE, M.J. (eds), pp. 458–92. Blackwell Scientific Publications, Oxford.
204. RAMUS, J., BEALE, S.I., MAUZERALL, D. and HOWARD, K.L. (1976). Changes in photosynthetic pigment concentration in seaweeds as a function of water depth. *Marine Biology*, **37**, 223–9.
205. RAMUS, J., LEMONS, F. and ZIMMERMAN, C. (1977). Adaptation of light-harvesting pigments to downwelling light and the consequent photosynthetic performance of the eulittoral rockweeds *Ascophyllum nodosum* and *Fucus vesiculosus*. *Marine Biology*, **42**, 293–303.
206. RAUNKIAER, C. (1934). *The Life-forms of Plants and Statistical Plant Geography*. Clarendon Press, Oxford.
207. RAYMONT, J.E.G. (1980). *Plankton and Productivity in the Oceans*, 2nd edn, Vol. 1, *Phytoplankton*. Pergamon Press, Oxford.
208. REES, D.A. (1972). Shapely polysaccharides. *Biochemical Journal*, **126**, 257–73.
209. REIMOLD, R.J. and QUEEN, W.H. (1974). *Ecology of Halophytes*. Academic Press, New York.
210. RHEE, C. and BRIGGS, W.R. (1977). Some responses of *Chondrus crispus* to light. I. Pigmentation changes in the natural habitat. *Botanical Gazette*, **138**, 123–8.
211. RHODES, R.G. (1970). Relation of temperature to development of the macrothallus of *Desmotrichum undulatum*. *Journal of Phycology*, **6**, 312–14.
212. RUENESS, J. and RUENESS, M. (1975). Genetic control of morphogenesis in two varieties of *Antithamnion plumula* (Rhodophyceae, Ceramiales). *Phycologia*, **14**, 81–5.
213. RUSSELL, G. (1967). The ecology of some free-living Ectocarpaceae. *Helgoländer Meeresuntersuchungen*, **15**, 155–62.
214. RUSSELL, G. (1972). Phytosociological studies on a two-zone shore. I. Basic pattern. *Journal of Ecology*, **60**, 539–45.
215. RUSSELL, G. (1973). The 'litus' line: a re-assessment. *Oikos*, **24**, 158–61.
216. RUSSELL, G. (1977). Vegetation on rocky shores at some North Irish Sea sites. *Journal of Ecology*, **65**, 485–95.
217. RUSSELL, G. (1978). Environment and form in the discrimination of taxa in brown algae. In *Modern Approaches to the Taxonomy of Red and Brown Algae*, IRVINE, D.E.G. and PRICE, J.H. (eds), pp. 339–69. Academic Press, London.
218. RUSSELL, G. (1979). Heavy receptacles in estuarine *Fucus vesiculosus*, L. *Estuarine and Coastal Marine Science*, **9**, 659–61.
219. RYTHER, J.H. (1969). Photosynthesis and fish production in the sea. *Science, N.Y.*, **166**, 72–6.
220. RYTHER, J.H., DeBOER, J.A. and LaPOINTE, B.E. (1979). Cultivation of seaweeds for hydrocolloids, waste treatment and biomass for energy conversion. *Proceedings of the International Seaweed Symposia*, **9**, 1–16.
221. SCHONBECK, M. and NORTON, T.A. (1978). Factors controlling the upper limits of fucoid algae on the shore. *Journal of Experimental Marine Biology and Ecology*, **31**, 303–13.
222. SCHONBECK, M.W. and NORTON, T.A. (1979). An investigation of drought avoidance in intertidal fucoid algae. *Botanica Marina*, **22**, 133–44.
223. SCHONBECK, M.W. and NORTON, T.A. (1979). Drought-hardening in the upper-shore seaweeds *Fucus spiralis* and *Pelvetia canaliculata*. *Journal of Ecology*, **67**, 687–96.
224. SCHONBECK, M. and NORTON, T.A. (1979). The effects of brief periodic submergence on intertidal fucoid algae. *Estuarine and Coastal Marine Science*, **8**, 205–11.
225. SCHONBECK, M.W. and NORTON, T.A. (1980). Factors controlling the lower limits of fucoid algae on the shore. *Journal of Experimental Marine Biology and Ecology*, **43**, 131–50.

226. SCHRÖTER, K. (1978). Asymmetrical jelly secretion of zygotes of *Pelvetia* and *Fucus*: an early polarization event. *Planta*, **140**, 69 – 73.
227. SEIP, K.L. (1980). A computational model for growth and harvesting of the marine alga *Ascophyllum nodosum*. *Ecological Modelling*, **8**, 189 – 99.
228. SETCHELL, W.A. (1915). The law of temperature connected with the distribution of the marine algae. *Annals of the Missouri Botanical Garden*, **2**, 287 – 305.
229. SHARP, J.H. (1977). Excretion of organic matter by marine phytoplankton: Do healthy cells do it? *Limnology and Oceanography*, **22**, 381 – 99.
230. SHIMURA, S. and FUJITA, Y. (1975). Changes in the activity of fucoxanthin-excited photosynthesis in the marine diatom *Phaeodactylum tricornutum* grown under different culture conditions. *Marine Biology*, **33**, 185 – 94.
231. SHUTER, B. (1979). A model of physiological adaptation in unicellular algae. *Journal of Theoretical Botany*, **78**, 519 – 52.
232. SIEBURTH, J.M. (1979). *Sea Microbes*. Oxford University Press, New York.
233. SKIRROW, G. (1975). The dissolved gases — carbon dioxide. In *Chemical Oceanography*, RILEY, J.P. and SKIRROW, G. (eds). Vol. 2, 2nd edn, pp. 1 – 192. Academic Press, London.
234. SMAYDA, T.J. (1958). Biogeographical studies of marine phytoplankton. *Oikos*, **9**, 158 – 91.
235. SOUTHWARD, A.J. and SOUTHWARD, E.C. (1978). Recolonization of rocky shores in Cornwall after use of toxic dispersants to clean up the *Torrey Canyon* spill. *Journal of the Fisheries Research Board of Canada*, **35**, 682 – 706.
236. SPENCER, C.P. (1975). The micronutrient elements. In *Chemical Oceanography*, RILEY, J.P. and SKIRROW, G. (eds), Vol. 2, 2nd edn, pp. 245 – 300. Academic Press, London.
237. STEEMANN NIELSEN, E. (1975). *Marine Photosynthesis*. Elsevier, Amsterdam.
238. STEPHENSON, T.A. and STEPHENSON, A. (1972). *Life between Tidemarks on Rocky Shores*. Freeman, San Francisco.
239. STEWART, W.D.P. (1974). *Algal Physiology and Biochemistry*. Blackwell Scientific Publications, Oxford.
240. STOCKTON, B. and EVANS, L.V. (1979). Characterization of alginates from *Laminaria digitata*. *British Phycological Journal*, **14**, 128.
241. STOSCH, H.A. von, and DREBES, G. (1964). Entwicklungsgeschichtliche Untersuchungen an zentrischen Diatomeen. IV. Die Planktondiatomee *Stephanopyxis turris* — ihre Behandlung und Entwicklungsgeschichte. *Helgoländer Meeresuntersuchungen*, **11**, 209 – 57.
242. SUNDENE, O. (1962). The implications of transplant and culture experiments on the growth and distribution of *Alaria esculenta*. *Nytt Magasin for Botanikk*, **9**, 155 – 74.
243. TAYLOR, W.R. (1964). The genus *Turbinaria* in eastern seas. *Journal of the Linnaean Society of London, Botany*, **58**, 475 – 90.
244. TERBORGH, J. and THIMANN, K.V. (1964). Interactions between daylength and light intensity in the growth and chlorophyll content of *Acetabularia crenulata*. *Planta*, **63**, 83 – 98.
245. TOLBERT, N.E. and OSMOND, C.B. (1976). *Photorespiration in Marine Plants*. CSIRO, Melbourne.
246. TOWLE, D.W. and PEARSE, J.S. (1973). Production of the giant kelp, *Macrocystis*, estimated by in situ incorporation of ^{14}C in polythene bags. *Limnology and Oceanography*, **18**, 155 – 9.
247. TURNER, M.F. (1979). Nutrition of some marine microalgae with special reference to vitamin requirements and utilization of nitrogen and carbon sources. *Journal of the Marine Biological Association of the United Kingdom*, **59**, 535 – 52.
248. VESK, M. and JEFFREY, S.W. (1977). Effects of blue-green light on photosynthetic pigments and chloroplast structure in unicellular marine algae from six classes. *Journal of Phycology*, **13**, 280 – 8.
249. VINCE-PRUE, D. (1975). *Photoperiodism in Plants*. McGraw-Hill, London.
250. WAALAND, J.R., WAALAND, S.D. and BATES, G. (1974). Chloroplast structure and

pigment composition in the red alga *Griffithsia pacifica*: regulation by light intensity. *Journal of Phycology*, **10**, 193 – 9.

251. WAALAND, S.D. (1975). Evidence for a species-specific cell fusion hormone in red algae. *Protoplasma*, **86**, 253 – 61.
252. WAALAND, S.D. and CLELAND, R. (1972). Development in the red alga, *Griffithsia pacifica*: control by internal and external factors. *Planta*, **105**, 196 – 204.
253. WAALAND, S.D. and CLELAND, R.E. (1974). Cell repair through cell fusion in the red alga *Griffithsia pacifica*. *Protoplasma*, **79**, 185 – 96.
254. WAALAND, S.D., NEHLSEN, W. and WAALAND, J.R. (1977). Phototropism in a red alga, *Griffithsia pacifica*. *Plant and Cell Physiology*, **18**, 603 – 12.
255. WAALAND, S.D. and WAALAND, J.R. (1975). Analysis of cell elongation in red algae by fluorescent labelling. *Planta*, **126**, 127 – 38.
256. WAISEL, Y. (1972). *Biography of Halophytes*. Academic Press, New York.
257. WERNER, D. (1971). Der Entwicklungscyclus mit Sexualphase bei der marinen Diatomee *Coscinodiscus asteromphalus*. III. Differenzierung und Spermatogenese. *Archiv für Mikrobiologie*, **80**, 134 – 46.
258. WHEELER, W.N., NEUSHUL, M. and WOESSNER, J.W. (1979). Marine agriculture: Progress and problems. *Experientia*, **35**, 433 – 5.
259. WHITTAKER, R.H. and LIKENS, G.E. (1975). The biosphere and man. In *Primary Productivity of the Biosphere*, LIETH, H. and WHITTAKER, R.H. (eds), pp. 305 – 28. Springer-Verlag, Berlin.
260. WIDDOWSON, T.B. (1971). A taxonomic revision of the genus *Alaria* Greville. *Syesis*, **4**, 11 – 49.
261. WILCOX, H.A. (1980). Ocean farming: prospects and problems. *Span*, **23**, 56 – 9.
262. WILLIAMS, P.J.LeB. and YENTSCH, C.S. (1976). An examination of photosynthetic production, excretion of photosynthetic products, and heterotrophic utilization of dissolved organic compounds with reference to results from a coastal subtropical sea. *Marine Biology*, **35**, 31 – 40.
263. WILTENS, J., SCHREIBER, U. and VIDAVER, W. (1978). Chlorophyll fluorescence induction: an indicator of photosynthetic activity in marine algae undergoing desiccation. *Canadian Journal of Botany*, **56**, 2787 – 94.
264. WISE, D.L., WENTWORTH, R.L. and KISPERT, R.G. (1977). Fuel gas production from selected biomass via anaerobic fermentation. In *Biological Solar Energy Conversion*, MITSUI, A., MIYACHI, S., SAN PIETRO, A. and TAMURA, S. (eds). pp. 411 – 26. Academic Press, New York.
265. WOMERSLEY, H.B.S. (1954). The species of *Macrocystis* with special reference to those on southern Australian coasts. *University of California Publications in Botany*, **27**, 109 – 32.
266. WOOLERY, M.L. and LEWIN, R.A. (1973). Influence of iodine on growth and development of the brown alga *Ectocarpus siliculosus* in axenic cultures. *Phycologia*, **12**, 131 – 8.
267. YAMADA, N. (1976). Current status and future prospects for harvesting and resource management of the agarophyte in Japan. *Journal of the Fisheries Research Board of Canada*, **33**, 1024 – 30.

Index

Pheromones 114–16
Philippines 34, 165–7
Phormidium 3, 6
Phosphate 32, 75, 166
Phospho-enol-pyruvate (PEP) carboxylase
  65
Phospho-enol-pyruvate (PEP) carboxykinase
  66
Phosphorus
  as plant nutrient 67, 70, 73, 76
  C:N:P ratio 31, 72–3, 171
  in sea water 30–2, 38, 72, 75, 171
Photic zone 9, 13, 18, 20, 24, 26, 52–3,
  82–3, 119, 172
Photoinhibition 47, 57, 86
Photomorphogenetic pigments 96, 111–12
Photon irradiance 13, 131–2
Photoperiodism 111–12
Photorespiration 65
Photosynthesis 43, 67, 112, 135, 176–7
  action spectra 49–52, 53, 57
  bacterial 170, 175
  biochemistry 63–6
  carbon sources 28–30, 63–5
  enhancement effect 45–6, 51–2
  in situ 26, 28–30, 31, 51–2, 80, 84–8
  light saturation 44, 53, 55, 62, 75–6, 78,
    103, 131–2
  maximum rate (P_{max}) 44, 59–61, 66, 127
  P vs I curves 44, 57–60, 75
  physiology 43–6
  relationship to growth 43, 55, 59, 61, 67,
    75–8
  vs CO_2 supply 26–30, 38, 63–5, 126
  vs desiccation 126–9
  vs irradiance 44, 47, 53–4, 58, 60, 75–7,
    79, 131–2
  vs pigments 46–60
  vs salinity 61–3
  vs temperature 44, 61, 79
Photosynthetic quotient 87
Photosynthetic pigments
  accessory pigments 2–4, 45–9, 51, 55–7,
    59
  composition 2–4, 46–7, 50–1, 52–7
  concentration 44, 53, 55–60
  extraction 47–8
  fluorescence 45, 48–9, 126
  light-harvesting components 48–9, 51,
    56–7
  P700 45, 48, 49
  Photosystems I + II 45–6, 47–8, 51, 126
  pigment-protein complexes 47–9
Photosynthetic unit (PSU) 58–60
Photosynthetically active radiation (PAR)
  13, 14, 18, 82
Phototropism 106–7
Phycobilin 2–4, 47, 48–9, 51, 56, 160

Phycobilisome 49, 56
Phycocyanin 2–4, 47, 49, 51, 54, 56
Phycoerythrin 2–4, 47, 49, 51, 54, 56
Phyllogigas 141
Phyllophora 132
Physiological adaptation
  chromatic 52–5, 77, 135
  intensity 59, 77, 131, 135
  temperature 61, 77, 131, 143–4
Phytochrome 111–12
Phytogeography 139, 149–55
  phytogeographic provinces 149
  phytogeographic regions 149–52, 154
Phytoplankton
  diseases 174
  distribution 10, 18, 20, 139, 141–4, 150,
    152–3
  excretion 77–8, 172
  growth 79, 80–3
  nutrient relations 30–1, 69, 71–5
  patchiness 74, 80
  photosynthesis 57–61, 64–5
  productivity 28–30, 85–8, 89–91, 156
  sampling problems 80–1, 86, 142
Phycomycetes 3, 174
Pigments
  photomorphogenetic 96, 111–12
  photosynthetic see Photosynthetic
    pigments
Plankton (see also Phytoplankton, Zooplankton) 7, 9, 10, 80, 139, 148, 153, 171–2
Planktoniella 141
Pleonosporium 103
Plocamium 132
Plurilocular sporangia 109–11
Polar latitudes
  phytogeography 142, 147, 149, 152
  phytoplankton 81, 83, 142–4, 147
  temperature 20–2
Polarity in plant cells 92, 96, 102
Polarized light 96
Pollution 132–3, 166, 172
Polysaccharide 5, 68, 77, 85, 98, 158–60,
  172
Polysiphonia 11, 68, 94, 154
Porphyra 6
  ecology 11, 54, 134
  harvesting 156–7, 165–6, 174
  photosynthesis 50–1, 53–4, 62, 64
  pigments 3, 50–1, 54
  reproduction 94, 110, 111–12, 165
Porphyridium 69
Postelsia 137–8
Potassium 24–5, 67–8, 157, 160
Predation 136–7
Productivity 43, 85–91, 119, 153, 156,
  171, 176
  estimates in sea 28–30, 89–91